Academia and the World Beyond

Christopher R. Madan
Editor

Academia and the World Beyond

Navigating Life after a PhD

Springer

Editor
Christopher R. Madan
School of Psychology
University of Nottingham
Nottingham, UK

ISBN 978-3-030-82605-5 ISBN 978-3-030-82606-2 (eBook)
https://doi.org/10.1007/978-3-030-82606-2

This Springer imprint is published by the registered company Springer Nature Switzerland AG
The registered company address is: Gewerbestrasse 11, 6330 Cham, Switzerland

For the PhD students unsure how to answer "What will you do after?" and the supervisors struggling to support those considering non-academic careers.

Acknowledgments

First and foremost, I need to wholeheartedly thank all of the contributors (again) for agreeing to be interviewed and sharing their experiences with us in this volume. Without their interest and support, there truly would not be a book for you to read now. It has also been a delight to get to know them better through their respective interviews, and I particularly appreciate their willingness to share personal challenges and realizations. Progress was slowed by the COVID-19 pandemic and related changes to teaching and caring responsibilities, but this book attests to the dedication of all contributors to share their hard-earned insights for the benefit of others.

Next, I would like to thank my friends and colleagues, as their encouragement was critical in motivating me to follow through with this project. The more I talked to people about this project, the more I was told that it was a much-needed resource.

Contents

Editor and Contributors

Editor

Christopher R. Madan School of Psychology, University of Nottingham, Nottingham, UK

Contributors

Kelly J. Arbeau Department of Psychology, Trinity Western University, Langley, BC, Canada

Gavin Buckingham Department of Sport and Health Sciences, University of Exeter, Exeter, UK

Alison Caldwell University of Chicago Medicine, Chicago, IL, USA

Neuro Transmissions, Chicago, IL, USA

K. Andrew DeSoto Association for Psychological Science, Washington, DC, USA

Jens Foell Department of Psychology, Florida State University, Tallahassee, FL, USA

maiLab, Heidelberg, Germany

Eiko I. Fried Department Clinical Psychology, Leiden University, Leiden, The Netherlands

Anastasia Greenberg Wilmer Cutler Pickering Hale and Dorr LLP, Boston, MA, USA

Jamie L. Hanson Department of Psychology, University of Pittsburgh, Pittsburgh, PA, USA

Learning Research & Development Center, University of Pittsburgh, Pittsburgh, PA, USA

Cleyde V. Helena ASC Group LLC, Centerton, AR, USA

Muireann Irish School of Psychology, The University of Sydney, Camperdown, New South Wales, Australia

Brain and Mind Centre, The University of Sydney, Camperdown, New South Wales, Australia

Jessica M. Karanian Department of Psychological and Brain Sciences, Fairfield University, Fairfield, CO, USA

Alice S. N. Kim Teaching and Learning Research in Action, Toronto, ON, Canada

Stavroula Kousta Nature Human Behaviour, Springer Nature, Nature Portfolio, London, UK

Elliot A. Ludvig Department of Psychology, University of Warwick, Coventry, UK

Joseph M. Moran Rapid7, Boston, MA, USA

Google, Cambridge, MA, USA

Aaron Moss CloudResearch, Queens, NY, USA

Maira Quintanilha Quali Q Inc., Edmonton, AB, Canada

Arfon M. Smith Space Telescope Science Institute, Baltimore, MD, USA

GitHub Inc., Edinburgh, UK

Lucina Q. Uddin Psychiatry and Biobehavioral Sciences, University of California Los Angeles, Los Angeles, CA, USA

Ana E. Van Gulick Figshare, London, UK

Simine Vazire Melbourne School of Psychological Science, University of Melbourne, Melbourne, Australia

Matthew B. Wall MRI Applications, Invicro London, London, UK

Faculty of Medicine, Imperial College London, London, UK

Clinical Psychopharmacology Unit, University College London, London, UK

A PhD Is a Journey, But Where Will It Lead?

Christopher R. Madan

Christopher R. Madan

Abstract A common question posed to PhD students from friends and family is: 'What will you do after?' But many students are too focused on the PhD itself and have not yet had a chance to sufficiently think about post-PhD life. Academic careers share many commonalities with many non-academic careers, with skills learned within academia being valuable in other career paths as well. To provide further context, here I provide background in my own training and motivations for developing this book. Several topics are discussed in multiple interviews and summarised here, in addition to an overview of the informational interview and suggestions on how the interviews can be used. I also provide insights on how academic life changes when transitioning to faculty and how the connectedness of modern Internet technologies can offset some downsides to this transition. This chapter ends with suggestions for additional resources to refer to and closing thoughts.

C. R. Madan (✉)
School of Psychology, University of Nottingham, Nottingham, UK
e-mail: christopher.madan@nottingham.ac.uk

C. R. Madan (ed.), *Academia and the World Beyond*,
https://doi.org/10.1007/978-3-030-82606-2_1

Contents

Introduction

While a PhD and the pursuit of advanced scientific training are useful in their own right, a PhD should not be viewed as a direct path to an academic position. Moreover, successful completion of a PhD is not the mastery of a singular skill; rather, it involves a mixture of field-specific training, scientific writing, data analysis and other skills. Importantly, these skills are relevant in a variety of contexts outside of academia—and a PhD should be considered as valuable training for a myriad of career paths. A PhD is a substantial accomplishment regardless of field and project nuances, and PhD students nearing the end of their degree often have been overwhelmed with the impending conclusion of this monumental task and associated procedures. As such, the question of 'what next?' has not been thoroughly assessed, and options beyond the academic career track have not been sufficiently explored.

An academic career is often associated with extensive flexibility and freedom. However, most who complete a PhD will not be able to find a permanent position within academia—regardless of desire and capability. The current system simply does not have capacity for all or most PhD students to become academic faculty. Even if an academic position is secured, if one is hoping to pursue a research-focused career, it is further important to acknowledge that grant funding is far from guaranteed and is sometimes required to maintain the position or for promotion. Though most postdoctoral research fellows are cognizant of their job prospects in academia (Woolston, 2020; Afonja et al., 2021; Council of Canadian Academies, 2021), everyone must find a suitable path for themselves.

For both PhD students and supervisors, the topic of non-academic careers can be difficult to begin to discuss. PhD students may feel lost, stressed by the impending completion of the dissertation and worried about disappointing their supervisor by considering a non-academic career. Supportive PhD supervisors may be unsure how to bring up the discussion and feel unqualified to provide the necessary advice about non-academic career paths that they themselves only superficially are aware of. In either case, my intention in making this book has been to facilitate these conversations and provide initial information about potential career paths in a format that is approachable to all involved.

This volume is a compilation of 22 interviews, all with people who have completed a PhD and are now in an academic or non-academic position. While it is unfortunate that there are not enough new academic positions created per year to be even close to the rate at which PhDs are awarded, ideally prospective PhD applicants should be aware of this early on. The perspective here is that this is not a problem to be solved and that PhD training is valuable for careers outside academia. As such, I consider 'non-academic' as more appropriate than 'alternative academic' or 'alt-ac', as these terms implicitly make the judgment that academia is the default path. Even this naming approach has limitations as there is such a diversity of positions outside of academia, but having a broad categorisation label is still useful.

In deciding who to interview, I focused more on those in non-academic positions because hearing more about these journeys is particularly beneficial for those completing a PhD as these are paths less discussed in traditional academic training. Even then, many of these careers are 'academia adjacent', as they involve regular interactions with academics—such as being a full-time journal editor, technical salesperson, or industry research scientist. Other non-academic positions may rely on skills learned in an academic setting but not necessarily involve a continued relationship with academia, such as some data scientist positions. For the academics I interviewed, I focused more on the job roles other than research because these are more distinct from the training that a PhD student has likely had (e.g. teaching, administrative/service responsibilities, broader considerations in running a lab, etc.).

About the Volume Editor and Interviewer

My name is Christopher Madan, and I am an assistant professor at the University of Nottingham in the United Kingdom. I study an array of topics, including memory, affect, decisions, aging and neuroimaging. Progressing from an undergraduate research assistant to a faculty position has taken me through academic systems in Canada, Germany, the United States and the United Kingdom. I care greatly about mentorship, trainee success and passing along what experiences I can. Apart from working with students in the research labs I have been in and now in my own growing lab, I am also on the Diversity, Outreach and Training committee of the Cognitive Neuroscience Society, and have been for several years–as well as having spoken on several conferences' career or trainee professional development panels. Additionally, I have written articles reflecting on research experiences and career choices (e.g. Madan & Chen, 2012; Madan & Teitge, 2013; Tsoi et al., 2018), guidance on doctoral supervision and studies (Madan, 2021, 2022), and guides for technical skills that I have learned along the way (e.g. Madan, 2014, 2015a, 2016). Even with my research interests, I have highlighted the parallels between research findings and their academic skill implications, such as using memory principles to improve scientific writing and teaching practices (Madan, 2015b; Van Hoof et al., 2021).

When I was a postdoctoral research fellow, I considered positions both within and outside of academia. Part of my process was to conduct informational interviews with people I knew in different positions to get a better understanding of what those roles entailed. Considering the high rates of PhDs leading to non-academic careers, it is important that current PhD students consider the full array of potential career paths. Through the benefits of experience and being active in the science Twitter community, I now know even more people in a variety of careers. For this volume, I contacted many of them to seek out their advice, experiences and reflections—to help give those currently finishing a PhD a starting point to understand the variety of potential futures available to them.

Informational Interviews

This book is a collection of informational interviews with people that have completed their PhD and then continued in either an academic position or another career path. In either case, they have been successful and have a multitude of insights to share with those who are currently finishing their PhD. Though continuing in academia may seem like the only option for a recent PhD graduate, a myriad of other career paths exist. Even if you do not pursue any of these other paths, some consideration and exploration of other paths are critical—you do not know what you might be missing out on.

For each chapter in this volume, I conducted an asynchronous semi-structured interview with each contributor, where we wrote together in a shared online document. I first provided a common set of questions (listed below) and further asked follow-up questions based on the responses. This narrative format was intended to make it easy and enjoyable for the contributors to share their experiences, as well as for readers to find the information engaging. All interviews were completed between May 2020 and May 2021.

Most contributors have completed their PhD in Psychology or Neuroscience. As such, their advice and perspectives are particularly relevant to students in either field. These contributors have taken a variety of career paths. In the domain of non-academic positions, these range from data science and science communication to patent law and sales of scientific research equipment. In addition to those interviewed in this book, I have friends and acquaintances that have pursued other career paths but were not interviewed here. Some of these individuals are now in relatively academia-proximate jobs, such as working as hospital research coordinator or within a government funding agency to set funding priorities, but others have continued into quite distinct careers—requiring further degrees—such as journalism or doing a Master's in Business Administration (MBA). Given the expertise and skills associated with psychology and neuroscience research, I know many that have continued their careers into behavioural or data science across a variety of industries, including marketing, sports, real estate, retail sales, police, banking, healthcare research, book publishing, drug development, software development, automotives and technology companies (e.g. Apple, Instagram, Uber, Twitter).

The Interview Questions

Below is a list of standard questions I considered when conducting the interviews in this volume, with consideration to the type of position held by the interviewee.

1. Can you introduce yourself and tell me a bit about your current position?
2. What was the focus of your PhD? *(mention when, where, department).*
3. As you were finishing your PhD, what were you thinking about your career plans?
4. How have your career plans changed as you have continued on to your current position?
5. Can you tell us a bit about what day-to-day life is like in your current position?
6. What do you like most about your work?
7. And what do you like least about your work?
8. How do you think having a PhD has helped you succeed in your current position?
9. If someone currently finishing their PhD was considering a position similar to yours, how might they decide if it would be a good fit?
10. A lot of people like academia because they feel it gives them an opportunity to work on a topic that they deeply care about. Do you think this is also true in your current position?
11. Another reason many like academia is that they feel it provides them with more freedom than they think they would get in other positions. How much freedom do you feel you have to work on what you think is interesting?
12. If someone was interested in pursuing a similar career path, what would you suggest they do to better prepare themselves?
13. Based on your journey, what advice or suggestions do you want to pass on to someone who is currently finishing their PhD?
14. Is there anything else you'd like to tell someone reading this interview?

What Should You Do with These Interviews?

Apart from just reading these interviews yourself, you could use them to facilitate group discussions. As one example, in my research group, I have the first lab meeting of each month designated as 'professional development'. Some topics include discussing what a faculty position job entails, various skills involved in academic research, the purpose of a dissertation and viva and considerations for finding suitable postdoctoral positions. For a few months, we had discussions where four to five students each read a different interview and described a summary of it, highlighting aspects they found interesting. This approach worked well in my lab, but individual reading and a reading group just among graduate students and postdocs are options as well, e.g. if there are not many graduate students in the same lab or there are no typically weekly lab meetings already.

As a PhD student, you should consider doing your own information interviews with people that recently completed a PhD or new faculty in your department. If you do not know people from more senior cohorts of your PhD program, you could contact recent former students of your department or ask faculty in your department for their contact information. People you have met at research conferences or on social media (e.g. Twitter) may also be relevant to interview yourself.

If you supervise PhD students, I recommend you discuss the topic of non-academic careers with them, perhaps using these interviews, to demonstrate that you are comfortable discussing these options and that the student is not on their own if they choose to explore a non-academic career path. If the student is interested in these careers, do what you can to facilitate their journey. Obviously, degree requirements need to be fulfilled, but if a student is interested in data science positions, give them an opportunity to try more complex analyses or have an analysis-only role on another project that is additional to their PhD work. (For instance, attempt to re-create an analysis result using open data; see Madan, 2014.) If a student is interested in science communication, ask them to write a lay summary of a relevant paper. Consider suggesting they write a perspective/opinion paper that provides some novel synthesis but also is intended for non-topic experts. Related opportunities include Massive Science Consortium (https://massivesci.com/consortium/) and *Frontiers for Young Minds*. Discuss with the student and translatable training can occur, still within the realms that you have expertise in as a supervisor, but also forming the basis of a non-academic portfolio for the student.

Overview of Interviews

There are many unique perspectives provided by those interviewed, but before you begin, I thought it would be useful to briefly summarise peoples' current careers and highlight broad topics that are discussed in multiple interviews.

Some of the interviews are with individuals currently in research-focused academic positions, including Gavin Buckingham, Eiko Fried, Jamie Hanson, Lucina Uddin, Muireann Irish, Elliot Ludvig and Simine Vazire. Two interviews are with individuals in teaching-focused academic positions, Jessica Karanian and Kelly Arbeau. Alice Kim and Maira Quintanilha both used their academic skills as the basis of starting their own companies and continue to work with researchers. Several others interviewed are also in academic adjacent positions: Stavroula Kousta is a full-time journal editor, Aaron Moss helps researchers conduct online studies, Andrew DeSoto works in government policy for a major psychology society, and Cleyde Helena works in scientific equipment sales. Ana Van Gulick and Arfon Smith support scientific research through industry positions, working in management positions related to data organisation—though their paths there differed markedly: Ana held an intermediate academic position at a university library; Arfon worked as a research software engineer for a time and then in a leadership position for a large technology-oriented scientific institute. Two interviewees, Jens Foell and Alison Caldwell, are in science communication positions. Joe Moran trained himself further in data science and user experience research and now does customer research. Anastasia Greenberg furthered her studies and is now an intellectual property lawyer, working with scientists as clients. Matthew Wall has found a position at the interface of academia and industry. Also worth noting, Elliot Ludvig and Muireann Irish both have experience in industry positions and returned to academia. A few have moved between universities after having secured faculty positions (Gavin Buckingham and Simine Vazire).

With respect to topics discussed, a few interviewees shared their experiences on the two-body problem of both individuals in a long-term relationship finding employment, including Jamie Hanson, Aaron Moss and Elliot Ludvig. Considerations about having children were discussed with Muireann Irish, Elliot Ludvig and Maira Quintanilha.

For those in non-academic positions, there are many reasons why people have pursued these career paths, but this is not always self-initiated. Several individuals who were invited to contribute to this volume had complicated views about transitioning to another career, having still hoped that an academic position would have worked out, now left jaded by the process. These individuals were not yet in a place where they felt able to provide advice to others. The interview with Cleyde Helena covers some of these topics, but I further recommend the Recovering Academic podcast (http://recoveringacademic.net) that she hosts along with two others.

Many interviews discussed considering the transferable skills you have developed as part of your PhD and that you should consider what you are passionate about.

Opportunities Just a Few Clicks Away

Doing research at the edge of current knowledge can sometimes feel lonely and isolating. It can be easy to forget the broader context (see Might, 2010), but it is also the focus of your work and can be particularly jarring when starting an independent position. As a PhD student or postdoctoral researcher, you work with others that have overlapping interests in a research group. In many cases, you have journal club discussions about recently published papers.

As an example of how things change, when starting a first independent academic position, there won't be others in your department with interests as overlapping, and there might even be only a few that are in your same subfield, e.g. cognitive psychology. Outside of the position itself, you may have moved to a university from far away and do not yet know many people. When in a more junior position, there often is support from new labmates and officemates, but this is not the case here. Moreover, now that you are in a faculty position and supervising trainees, it is important to ensure you have boundaries that go along with the responsibilities of supervision and mentorship. For instance, frequently going out drinking with a subset of the lab could readily lead to the impression of favouritism, if not worse. While these are often the circumstances associated with starting a faculty position, the world is changing, academia along with everything else. In particular, the availability of the Internet can do a lot to broaden who we can connect with.

Applying for jobs itself is a lot less stressful if you have friends who are also at the same career stage who are also applying. Even more so when you each find positions you can collaborate with each other and figure things out together. While I collaborate with a few people, for me, the key friend here is Daniela Palombo. We were both postdocs in Boston, though at different universities, and soon found that we have a lot of shared research interests but have been using different approaches. Both of us are faculty now, and even though we are eight time zones apart, we chat often and discuss research ideas as well as other topics that have come with the role, such as teaching and approaches to graduate student supervision. We also commiserate with each other about challenges, such as less favourable teaching evaluation comments and grant rejections. Quite a few collaborative projects have come from these frequent conversations, and working with her has definitely been important for me to feel like I have someone to talk to as I learn to navigate these new dimensions of academia. Although this example pertains to my experience in an academic position, having close friends—even time zones away—is important to maintain, especially if repetitively uprooting your life every few years for the next position. Of course, it is also important to engage in your department and make new friends where you are. Go for coffee with people who have newly started and attend departmental social events (try to arrive with someone you know!). Try to present your recent work at an internal seminar to give others more insight into what you do now that the time pressure and stress of the interview have long passed and you are a colleague.

Social media can be a great resource for keeping connected to those far away and making new connections based on other shared circumstances, such as being new faculty (Madan, 2017; Cheplygina et al., 2020; Foell, 2021). This can be particularly useful for students to share their triumphs and challenges and to reduce imposter syndrome.

Additional Resources

The goal of this book was to collect and then share personal experiences and lessons learned from many who have progressed beyond a PhD and been successful. This interview approach to providing PhD-related advice is unique to this book, but the discussion of career considerations—both academic and non-academic—is an important topic discussed in many resources. For broad advice about choosing an ideal career track, see Bielczyk (2020). For academic advice, including advice with job applications and grant writing, see Kelsky (2015), Barker (2010), Gabrys and Langdale (2012), or Feibelman (2011). For non-academic advice, I suggest readers to see Caterine (2020), Linder et al. (2020) and Duckles et al. (2020). For data science careers specifically, see Robinson and Nolis (2020). Do consider if these or related resources are available through your university library.

Final Observations

Beginning a PhD does not mean you are expected to commit to academia for life. Both academic and non-academic careers are associated with a long list of relevant skills and, depending on the specifics, bear a lot of commonalities (Vitae, 2010; Edge & Munro, 2015; Wright & Vanderford, 2017; Weber et al., 2018; Duckles et al., 2020). Some of these include creative problem solving, writing, technical skills, leadership abilities and public speaking. Academic positions themselves have a diversity and variability in responsibilities, much more than just research. Moreover, there are many non-academic positions that strongly benefit from PhD training—a PhD involves many skills; consider what are your strengths and what you enjoy and use these to choose a suitable career path. If you do explore non-academic career paths further, be considerate of how to present your skills professionally within these contexts, to use their language and reframe your work experience and skills to do justice to your years of experience, but within a non-academic context.

I sincerely hope you find these interviews helpful in considering and exploring potential career paths. It was a joy to get to hear more about the lives of those interviewed and help share their experiences, insights and advice.

References

Afonja, S., Salmon, D. G., Quailey, S. I., & Lambert, W. M. (2021). Postdocs' advice on pursuing a research career in academia: A qualitative analysis of free-text survey responses. *PLoS One, 16*, e0250662. https://doi.org/10.1371/journal.pone.0250662

Barker, K. (2010). *At the helm: Leading your laboratory* (2nd ed.). Harbor Press.

Bielczyk, N. (2020). *What is out there for me? The landscape of post-PhD career tracks*. Welcome Solutions.

Caterine, C. L. (2020). *Leaving Academia: A practical guide*. Princeton University Press.

Cheplygina, V., Hermans, F., Albers, C., Bielczyk, N., & Smeets, I. (2020). Ten simple rules for getting started on Twitter as a scientist. *PLoS Computational Biology, 16*, e1007513. https://doi.org/10.1371/journal.pcbi.1007513

Council of Canadian Academies. (2021). *Degrees of success*. The Expert Panel on the Labour Market Transition of PhD Graduates, Council of Canadian Academies. https://cca-reports.ca/reports/the-labour-market-transition-of-phd-graduates/

Duckles, B. M., & Open Post Academics Contributors. (2020). *Leaving the academy: Resources from Open Post Academics*. http://book.openpostac.org

Edge, J., & Munro, D. (2015). *Inside and outside of the academy: Valuing and preparing PhDs for careers*. Conference Board of Canada. https://www.conferenceboard.ca/edu/research/valuing-preparing-phds-for-careers

Feibelman, P. J. (2011). *A PhD is not enough! A guide to survival science*. Basic Books.
Foell, J. (2021). Social media science communication is a nonstop academic conference for all. *Nature Humam Behaviour*. https://doi.org/10.1038/s41562-021-01138-0
Gabrys, B. J., & Langdale, J. A. (2012). *How to succeed as a scientist: From postdoc to professor*. Cambridge University Press.
Kelsky, K. (2015). *The professor is in: The essential guide to turning your Ph.D. into a job*. Three Rivers Press.
Linder, K. E., Kelly, K., & Tobin, T. J. (2020). *Going alt-ac: A guide to alternative academic careers*. Stylus Publishing.
Madan, C. R. (2014). An introduction to MATLAB for behavioral researchers. *Sage*. https://doi.org/10.4135/9781506335131
Madan, C. R. (2015a). Creating 3D visualizations of MRI data: A brief guide. *F1000Research, 4*, 466. https://doi.org/10.12688/f1000research.6838.1
Madan, C. R. (2015b). Every scientist is a memory researcher: Suggestions for making research more memorable. *F1000Research, 4*, 19. https://doi.org/10.12688/f1000research.6053.1
Madan, C. R. (2016). Improved understanding of brain morphology through 3D printing: A brief guide. *Research Ideas and Outcomes, 2*, e10398. https://doi.org/10.3897/rio.2.e10398
Madan, C. R. (2017). Beyond the ivory tower: Reciprocal relationships between scientific research, communication, and mentorship. Oral presentation given at the *Annual Convention of the Association for Psychological Science*. https://doi.org/10.7490/f1000research.1114114.1.
Madan, C. R. (2021). A brief primer on the PhD supervision relationship. *European Journal of Neuroscience, 54*(4), 5229–5234. https://doi.org/10.1111/ejn.15396
Madan, C. R. (2022). Approaching the PhD thesis and viva: Clarifying expectations and enhancing preparations. *Cognitive Psychology Bulletin, 7*, 46–48.
Madan, C. R., & Chen, Y. Y. (2012). Getting involved in research: A guide for inspired undergraduates. *Eureka, 3*, 2–4. https://doi.org/10.29173/eureka16983
Madan, C. R., & Teitge, B. D. (2013). The benefits of undergraduate research: The student's perspective. *The Mentor, 15*, 61274. https://doi.org/10.26209/MJ1561274
Might, M. (2010). *The illustrated guide to a Ph.D.* Retrieved from http://matt.might.net/articles/phd-school-in-pictures/
Robinson, E., & Nolis, J. (2020). *Building a career in data science*. Manning Publications.
Tsoi, M., Teitge, B. D., Madan, C. R., & Francescutti, L. H. (2018). Personal values influencing career path in academic medicine: Perspectives of selected Canadian trainees. *F1000Research, 5*, 1903. https://doi.org/10.12688/f1000research.9026.2
Van Hoof, T. J., Sumeracki, M. A., & Madan, C. R. (2021). Science of learning strategy series: Article 1, distributed practice. *Journal of Continuing Education in the Health Professions, 41*, 59–62. https://doi.org/10.1097/CEH.0000000000000315
Vitae. (2010). *Researcher development framework.* Careers Research and Advisory Centre. http://www.vitae.ac.uk/rdf
Weber, C. T. et al. (2018). *Identifying transferable skills and competencies to enhance early-career researchers employability and competitiveness*. European Council of Doctoral Candidates and Junior Researchers. https://doi.org/10.5281/zenodo.1299178.
Woolston, C. (2020). Uncertain prospects for postdoctoral researchers. *Nature, 588*(7836), 181–184. https://doi.org/10.1038/d41586-020-03381-3
Wright, C. B., & Vanderford, N. L. (2017). What faculty hiring committees want. *Nature Biotechnology, 35*, 885–887. https://doi.org/10.1038/nbt.3962

"A Career Path in Open Science Has Been a Great Fit"

Ana E. Van Gulick

Ana E. Van Gulick

Abstract In our interview with Ana Van Gulick, she tells us about her PhD work and struggles with working alone and the slow pace of research. She subsequently worked as an academic librarian and learned more about data management and open science. Ana now works in a project manager role for Figshare, a company that serves as a scientific data repository. This role allows Ana to help researchers share their data, keep up with changing scientific research practices, and work across many disciplines. A PhD prepares people for many career paths beyond the specific academic field of study, as many skills are transferable.

Contents

A. E. Van Gulick (✉)
Figshare, London, UK
e-mail: ana@figshare.com

C. R. Madan (ed.), *Academia and the World Beyond*,
https://doi.org/10.1007/978-3-030-82606-2_2

Chris: Can you introduce yourself and tell me a bit about your current position?

Ana: Hi! I'm Ana. I was trained as a cognitive neuroscientist and currently work on supporting open research data at a technology company called Figshare that hosts a generalist data repository and builds data repository infrastructure for organizations. In my role as the Figshare Government and Funder Lead, I manage projects for clients in our government and funder sector that are both public agencies and private non-profits. I support both repository administrators and researchers who are the repository end users in implementing data sharing best practices so research results are discoverable and reusable. I also manage reporting and contract deliverables for these projects and work with our marketing and sales teams to find opportunities for our infrastructure to support open science at research organizations.

In my undergraduate research and for my PhD, I studied the high level visual system with both behavioral and MRI methods. I investigated how we recognize and categorize objects that we have a lot of experience with, such as faces. After completing my PhD, I worked in an academic library for 6 years. At Carnegie Mellon University (CMU) in Pittsburgh (Pennsylvania, USA), I first joined the University Libraries as a postdoc as part of a fellowship program run through Council on Library and Information Resources (CLIR) that brings PhD researchers into academic libraries in an effort to expand how academic libraries are evolving to respond to the changing needs of researchers. After nearly 2 years of postdoc, I was hired as faculty at the libraries to serve as the Liaison Librarian to Psychology and Neuroscience. The Librarian track at CMU is one without tenure, but with benefits and academic review similar to a teaching or research track. In both my postdoc and faculty role at the libraries, I focused much of my effort on open data practices and open science research workflows. I founded and directed the CMU Libraries Open Science Program that included training workshops, digital tools, and expert consultants to support more transparent, reproducible, and reusable research across disciplines.

What was the focus of your PhD?

My PhD training and research were at Vanderbilt University (Nashville, Tennessee, USA) in the Department of Psychology. There I was focused on understanding how we recognize and categorize objects using cognitive neuroscience approaches. To study these high-level visual processes, that is what happens following processing in the eye and primary visual cortex, we used both behavioral and neuroimaging techniques such as functional MRI to study downstream processing of visual objects. One question my PhD lab was particularly interested in studying was the effect of experience on visual recognition and a phenomenon called "perceptual expertise." Both the amount of experience and the type of experience recognizing objects seem to influence how objects are recognized and what brain areas are recruited for this. There are certain categories of objects in the world such as faces, with which we have a high level of experience. Face recognition also requires not just categorization at a broad level, but individuation to recognize a specific person, which is a challenging visual task given how similar faces are to one another. It's been found that face recognition has certain attributes such as more holistic processing and recruits specific brain areas relative to recognition of other object types.

Much of my PhD work was aimed at understanding why these differences are observed with objects like faces by working to understand the effects of different types of experience. In some studies, this was done with training studies in which participants spent 10–-20 hours learning to categorize different novel computer-generated objects at the individual and higher category levels to see what impact these types of experience had on visual recognition. In other studies, we used real-world experience as the variable such as working with car and bird experts, that is, people with a personal interest in birds or cars who have both significant knowledge and experience with these categories. By designing visual and semantic tests about a variety of object categories from planes to shoes to mushrooms, we tried to parse the contribution of different types of experience versus baseline visual and intellectual skills to understand how perceptual expertise develops across individual differences in people.

Overall, we wanted to understand how experience affects visual recognition across individuals and how these differences are observed in the neural bases of the high-level visual system in the brain.

As you were finishing your PhD, what were you thinking about your career plans?

Over the course of the 5 years of my PhD program, I became increasingly interested in pursuing a career outside of the typical faculty tenure track. One of the main reasons for this realization was learning what I enjoyed about doing research and what I didn't. By my third or fourth year of graduate school, I had found that I really enjoyed active and collaborative work, whether it was working on research projects with other students and postdocs in the lab, doing analyses together with a colleague, or having subjects visit the lab and collecting data. This type of work made me feel like I had accomplished something each day and kept me motivated to do research. What I found I struggled with more was working alone, such as long periods spent programming code, reading the literature, or writing proposals or papers alone.

After having a first-year research project that moved slowly and wasn't published until my fifth year of graduate school, I learned about the sometimes truly slow pace of scientific research. While I think I had some understanding that research takes time from my undergraduate experience, I found it challenging to have projects that took not just many months or a year but truly took many years. A faster pace of accomplishment like that more common in non-academic jobs would probably suit me better. I also did very little teaching as a graduate student as I was fortunate to have other sources of funding for my research and training, but based on my limited experience, I did not think that I wanted to have a future position with a heavy teaching component.

Realizing that the "active" part of research was what I enjoyed most and that faculty members spend little time doing research and much more time teaching and writing grants and papers, it seemed that wouldn't be the right match for me. Of course, that's to say nothing of the competitiveness of the tenure track job market, which would also have been a challenge.

Seeing that I liked working on a team and having projects that moved faster, it seemed that a non-academic position in a research field or applying scientific processes and knowledge in an industry setting would be a good path forward for me. Toward the end of my PhD, I was thinking about applying my research skills to scientific consulting or to user experience in the commercial sector.

To be honest, working in open data and open science was not a career path I considered in graduate school, although this was also a newer movement in science at that time as open science has evolved rapidly in the past 7 years. I also never considered working in an academic library on these issues as my experience with the library as a student was strictly limited to downloading journal articles online. I can say now though that a career path in open science has been a great fit – it allows me to be part of the research ecosystem, to support researchers and scientific discovery, and it is an exciting and quickly evolving field.

Working in an academic library was a great pathway into the fields of research data management and open science. This is a new area of work for academic libraries, and I feel like I got to have a significant impact on the program at CMU and to contribute to the broader library and open science fields. Nonetheless, the professional path forward in libraries usually involves becoming an administrator, such as an associate dean or university librarian and part of the library leadership, and after several years as library faculty, I decided I preferred to focus on working exclusively on open data and research rather than having to broaden my interests into more general academic library administration.

At Figshare, which is a relatively young (10 years) technology company, I've been able to focus on open research and work in a fast-paced environment focused on data repository infrastructure and support with an energetic and dedicated team.

Can you tell us a bit more about how you went from finishing your PhD to working in an academic library? Did you need to do any formal training in advance?

After completing my PhD, I began my postdoc in an academic library (at Carnegie Mellon University, Pittsburgh, PA, USA) with no formal training and really no idea whatsoever about the field I was entering. It was in many ways (and as with so many life stories) quite a matter of chance. The CMU Libraries had applied to host a CLIR postdoc focused on data in the sciences and social sciences but had had trouble recruiting someone from a science background for the position. The humanities reputation of libraries, I learned later, often results in this recruiting issue, despite the exciting work on data that libraries are doing currently. I was looking to move to Pittsburgh to be with my partner there, and a former advisor of mine, who was faculty at the university, pointed me in the direction of the position. I was pretty hesitant the position would be a good fit, given that I had visited the library in person exactly once during my 5 years as a PhD student, but seeking a first job to make the move, I was interviewed and was offered the job the morning after my dissertation defense.

Because the job was part of a larger postdoctoral fellowship program through CLIR, there was some training provided. I attended a 10-day training with the other 25 fellows in my cohort, all recent PhDs, which was aimed at giving us a crash course in academic libraries. Like so many disciplines, the library world is full of acronyms and jargon, and I will admit to feeling confused and fully out of my element at the training. I was one of only a few scientists in the cohort as many postdocs came from digital humanities backgrounds, which was a focus of some of the positions that year, which also contributed to my "fish out of water" feeling. At the same time though, many similarly had no library experience, and having this cohort was critical to helping me navigate the first year of my postdoc. We had monthly calls and met up at conferences and worked through the challenges of being the odd recent PhD working in a library. We faced similar hurdles in finding projects that we could contribute to and work that was appropriate for our skills while also integrating ourselves into the library culture.

I think culture is really the biggest part of what I needed to learn to work in a library actually. Researchers and librarians often talk about the same things but use different words. As a postdoc and later as library faculty, I felt that my role was often one of translator and then connector. So once I learned the "library language," I could interpret both ways – make library resources approachable to researchers by speaking their language and in the reverse help those in the library understand the practices of researchers.

That was the extent of the formal training, but I did learn on the job from those in the library such as working with our data and scholarly communication librarians. I was also lucky to have significant support to attend workshops and conferences, which was a huge help to understanding the larger research data management and open science communities. These data and open science communities have grown significantly in the past 6 years and are now hopefully more accessible to those working in research communities including graduate students than they were when I was a student. Some events that I attended were focused on the library community, often data librarianship, a relatively new subdiscipline, and others brought together librarians, researchers and those from funders, publishers, technology companies, governments, and nonprofits as there are many stakeholders in the research ecosystem. The later I found a quite exciting community, and it is the world I have continued to work in following my time in libraries.

Can you tell us a bit about what day-to-day life is like in your current position?

The best description of my current position is probably as a "project manager," but Figshare is fairly small and still has a scrappy start-up culture, so I do a bit of a lot of things. My main responsibility is managing projects for our clients who are governments or funders. As a technology company, Figshare builds the infrastructure for open access research repositories. We provide Figshare.com as a free, generalist repository that researchers around the world can use to share any research products, and we also provide our infrastructure as a subscription to our clients to build custom repositories that meet the needs of their specific institution or their researchers. In my first year at Figshare, I have primarily supported US government agencies who are looking to comply with mandates to provide public access to data as well as to support both their intramural and extramural researchers with a repository that can be used to make the results of their work discoverable, reusable, and more impactful. For these projects, I serve as the primary Figshare point of contact for both the project leads on the client side and for questions and support for the end users, that is, the researchers using the repository. From the contracting side, I provide regular reports on the project progress and repository usage and impact as required by the contract and meet any other contract deliverables. I conduct outreach and training such as webinars for researchers and even work with the academic librarian community who can point their researchers to some of these repository resources if they are funded by one of the agencies we work with. I also do curation of deposited research for some repositories.

Working with researchers on sharing data is something I did at the CMU library as well, which has an institutional repository powered by Figshare. For some repositories, especially in the government and funder space, Figshare is contracted to provide not only the infrastructure but also the experts to support researchers and review deposits before they are published. The goals of this review are to check the files and ensure the work is appropriate for the repository but most importantly to check the documentation included with the files and the metadata that will make the work discoverable in search engines and reusable by other researchers. At the beginning of open science, the emphasis was on making the results of research open, but now as the open data movement evolves, there is increasing recognition that open does not always equal Findable, Accessible, Interoperable, or Reusable (FAIR, a common acronym in the field). Part of my role at Figshare is thinking strategically and working with clients and the larger repository and open science community on how we can enhance open data by applying emerging standards and best practices. Having an expert lead researchers through the data sharing process and check that metadata is as complete as possible is one way we're encouraging this.

In one project, I got to work on a pilot with the National Institutes of Health who partnered with Figshare on a 1-year pilot project to host a generalist repository for NIH-funded researchers. Some of the goals of the project were to evaluate the need for a flexible generalist repository in the biomedical data landscape and also to determine the effect of having a data curation expert guide researchers in using the repository and check datasets and other deposits before they were published publicly. Working with NIH-funded researchers, both intramural researchers employed by the NIH and extramural researchers at academic institutions who receive NIH grants, really brought together all of my career experiences bridging my experience as a scientific researcher with NIH funding (funding that had supported a large portion of my PhD training) and as a librarian working with open data practices. Now working with a large funder like NIH, I had the chance to provide a repository resource to a large audience of researchers across biomedical fields and to participate in the NIH open data ecosystem of discipline-specific repositories, generalist repositories, researchers, and specific NIH institutes funding the research.

How do you think having a PhD has helped you succeed in your current position?

Most importantly, I think it comes back to culture and understanding what the workflows, practices, and day-to-day activities of an academic researcher are. Being a PhD student exposed me to the reality of scientific research including the many stages of research project from planning and data gathering through data analysis and visualization, the ordeals of writing (and only sometimes getting) grants for funding, the often lengthy process of publishing papers in peer-reviewed journals, and many other institutional and lab workflows. I had grown up with a parent who is an academic, so I know quite a lot about universities having grown up around graduate students and listened to my father both teach and navigate university politics. But my father is a philosopher and so there is quite a difference in the day-to-day for a researcher in the sciences as well as a dramatic difference in the funding and publishing landscape. In my current position, understanding the life cycle of research data and the workflows for data and publications is key to understanding researchers needs and supporting them. Additionally, understanding the grants process and requirements of large funders such as the National Institutes of Health and the National Science Foundation especially around issues such as data management and public access has been a great advantage for my current position. While I work in tech for a private company now, the larger community that makes up my clients, colleagues, and users is still the same one I was a part of as a PhD student.

What do you like most about your work?

Working in open research has been a great fit for me because I get to work with researchers and funders across many disciplines and help shape the future of scientific discovery, While I didn't want to continue on a tenure-track path, I really enjoyed working in cognitive neuroscience and being a part of the scientific research community. In my current position, I get support to do primary research from the infrastructure and training side and be a part of the larger research community still. It's been particularly exciting working with government agencies and other funders who support this work, as they are such an important driver of research practices and honestly allow research to happen. Scientific research is changing quickly with the rise of big data, digital data, powerful computing, and artificial intelligence, and getting to be a part of that through a product that I can help shape and projects that I manage is a great opportunity to support research reproducibility and discovery. I get to work with a great group of motivated colleagues and be part of professional networks that span the globe, and I think we will have a meaningful impact on what open data looks like in 10 years. I also enjoy managing complex projects (lots of to-do lists and post-it notes!) and working with lots of different people and different projects, so that diversity of experience in my day-to-day is great. While any labs I worked in as a PhD student were usually funded by the National Eye Institute or occasionally National Institute of Mental Health if they had NIH funding, in this position, I've gotten to look at data and work with researchers funded by more than 20 other NIH institutes across biomedical fields. I certainly don't always know what the data are or what they mean, but it's quite interesting to observe and learn about some many different types of research being done.

And what do you like least about your work?

The paperwork and reporting required for grants and contracts are probably high on my list of things I could stand to do less of as there is *a lot* of it as a government contractor and project manager. But I understand why it's needed. There is also a sales element to my work that can be stressful if it's not going well – while I'm not responsible for any sales quotas myself, I work with members of our sales team on finding opportunities and presenting our products to prospective clients. That being said, it is really great when you find someone with a problem that your product solves perfectly, and I really believe in the mission of the company, which is dedicated to making research openly available, so it's easy for me to stand behind our work.

I would also mention what I have experienced as a darker side of the open research community – one that can be quite exclusionary or antagonistic and seems to conflict with the mission of the movement. From my perspective, I've seen this take several forms including a strong in-group bias that's not welcoming to newcomers or those from diverse backgrounds, an overemphasis on computational or programming skills, and a combative stance toward only endorsing open source versus commercial or licensed solutions. There are many great people working in this community who are aware of these issues and actively working to resolve them especially around diversity and inclusion. Similarly, there are undoubtedly commercial players in the academic space that are a detriment to research, but I hope sustainability models, mission, and community contribution can also be considered rather than drawing a black or white distinction between open source or nonprofit solutions and those that use a commercial model to be sustainable.

A lot of people like academia because they feel it gives them an opportunity to work on a topic that they deeply care about. Do you think this is also true in your current position?

Getting to work on something you care deeply about is certainly a common refrain among academics; however, I think whether this is true for you in academia or in any industry is very individual and not necessarily tied to the type of employer. I was very passionate about working to understand human vision and cognition as a student and still am interested in this field, but I'm not sure that I was ever that passionate about a specific study I was working on. In many ways, I think I am more passionate about the work I do now because it is broader and easier to see the impact day to day. In academia, research areas become quite narrow quickly, so the work that you are doing is often focused on very granular problems and nuanced theories. In my experience in academia, it was hard to see the forest when you were working so hard on the trees in this scenario. In my non-academic roles in libraries and at a data repository, I've gotten to work on something I've come to care deeply about, open and reproducible research, in a way that more frequently includes a broader view, which keeps me motivated in the day-to-day work of projects and details. Even in academia, where you can choose your research program, you may not get to choose all of your work, for example, what courses you teach or what committees you serve on. I've found this to be similar in non-academic jobs as well, but luckily in my case, I personally care about the mission of the organization, which has led to a high level of personal interest in my work. Overall for me, I don't feel that I've sacrificed being passionate about what I work on for my non-academic career because I was able to find other work that I was equally interested in doing.

Another reason many like academia is that they feel it provides them with more freedom than they think they would get in other positions. How much freedom do you feel you have to work on what you think is interesting?

Certainly, there are few jobs with as much freedom to choose your own work as academia, and many that do offer that freedom (e.g., being an artist or a writer) don't offer the same support and security of working at an academic institution. I've been very lucky in both of my post-PhD jobs, one at an academic library and one at a company, to have supervisors that provided a great deal of intellectual freedom to pursue new projects and to do my job as I saw best. In my current role at a data repository, I have work that must be done for which I don't have the freedom to choose which projects or tasks to work on as they are by nature part of my job. I would say this quite "well-defined" work makes up roughly half of my work and the other half is more flexible for me to design on my own even as it contributes to the goals of my job description. In this way, I can choose to do a webinar, write a guide or blog post, attend a workshop, or even come up with ideas for business strategies and product features on my own and contribute to my job, but in a way that is not prescribed such that it is still creative work. And while my main role takes much of my time, I'm still encouraged to continue professional development and training and can pursue my own research and scholarly work and be involved with research community activities as I have the interest and time. At Figshare, our team is still fairly small, and the relatively new business and products are still evolving quickly, which I think has given me more of a chance to take an active role contributing creatively to my job versus doing predefined work all the time. I'm also actively engaged in evaluating my job description regularly and have been encouraged to continue to adjust and refine, which certainly has to be done with business needs in mind as well as my own interests, but I am grateful for this flexibility. In the future, I hope this will lead to opportunities to be a leader at the organization.

I certainly don't have as much freedom as I would working in academia, but I have also been quite happy with the balance of freedom I do have. Recently, my supervisor asked if I was okay with the "lack of structure" in my day-to-day work, his thinking being that as a newer technology company, we work with quite a lot of uncertainty and still have a start-up-like organization structure, to which I responded that there was still vastly more structure than in academia, so I was quite happy with it. Like many PhDs, I'm pretty self-motivated to do my work and enjoy doing some work that is creative, but I also know that I prefer to work in a pair or small group and to have some degree of structure to my work goals. Given that I wouldn't want 100% freedom anyway (that sounds like a paralyzing lack of structure and teamwork), having the opportunity to contribute creatively to the organization and how I achieve my individual goals at work as well as the opportunity to shape my role in the future seems to be the right balance for me.

What are some relative benefits you've seen in working in a non-academic position?

I think that a common fear among those considering leaving academia is leaving the culture and community of academia. I grew up with an academic parent and had been around universities my entire life through graduate school and also experienced this concern in considering alternative careers, so it seems worth noting my experience on the other side. Academic culture has many wonderful qualities including flexibility and freedom, institutional support and benefits, intelligent and curious colleagues, and a continuous stream of engaging lectures and events to attend. However, as has been increasingly noted in recent years, academia is also highly competitive, which can lead to a culture of overwork and a lack of work/life balance despite the flexibility and can also reinforce biases in who succeeds. Depending on your personality and your approach to work, other academic adjacent career paths might offer structure and balance that are beneficial with less stress than academia.

In my roles outside of academic research, both at an academic library and currently at a company, I've found improved work/life balance including more respect for "off work hours" as well as many of the attributes I enjoyed about academic culture. In my current role, I rarely receive work messages on evenings or weekends, and if I do, I don't feel pressured to reply. Similarly, vacation time is respected as a time you can fully disconnect, which I find very valuable as well as having a dedicated vacation time allotment, which I've luckily found to be quite generous by American standards. At my current job, I still have the flexibility to design my own work day around meetings and commitments, to step out for a doctor's appointment or even just to take a break for a walk and shift my work time when needed – something I feared might vanish in an "industry job." At both jobs, I've found smart, highly motivated colleagues who are supportive and engaged collaborators, and some days I'm in meetings with just as many PhDs as if I worked in academia. The work culture of any organization should always be an important consideration for any job, and considering the pros and cons of academic culture should be no different; find a culture that is the right balance for you to be successful and happy whether that's in academia or elsewhere.

Based on your journey, what advice or suggestions do you want to pass on to someone who's currently finishing their PhD?

I would tell them that there are many opportunities both within academia (in traditional faculty roles and nontraditional roles) and in many other industries. They should feel confident that their experience from earning a PhD has prepared them to do many jobs well and that many of these jobs may seem far from their field of study or distant from their real-world experience, but that does not mean they would not be happy and successful doing them. Think about the type and style of work you find enjoyable when choosing a career as well as the topic of your expertise. Many skills you develop as a PhD candidate – writing, synthesizing information, critical thinking, and problem-solving – are applicable across many domains. Good advice seems to be to do informational interviews with people working in different fields to learn more about them and how you might apply for jobs in that field if you are interested, although to be honest I did not do this myself (however, I have spoken to numerous PhD students since leaving academic research about my jobs!). When choosing a post-PhD career, feel free to consider things like work/life balance, work culture, salary, and location as considerations if they are important to you – they may even be more important to you than the specific projects you work on.

Is there anything else you'd like to tell someone reading this interview?

I think that the growth of remote work in many including tech may significantly broaden the non-academic options for careers post-PhD and hope this will help more PhDs find jobs that are the right fit for them.

Thank you so much for sharing your experiences with us, Ana!

"Technical Skills Are Always Useful, No Matter What You're Actually Working On"

Matthew B. Wall

Matthew B. Wall

Abstract In our interview with Matthew Wall, we learn about the potential to work in a position that bridges between commercial and academic research. Using the cognitive neuroscience approach of functional magnetic resonance imaging (fMRI), Matt conducts research on the effects of drugs on the brain. After a difficult PhD studying implicit learning and resulting in no publications, technical experience with fMRI and programming acquired during his PhD laid the foundation for his career. While it is useful to have a career plan, you also need to be flexible and realistic about your options. Strong research skills and an interest in methods are valuable and could lead to a path like Matt's.

Contents

M. B. Wall (✉)
MRI Applications, Invicro London, London, UK

Faculty of Medicine, Imperial College London, London, UK

Clinical Psychopharmacology Unit, University College London, London, UK
e-mail: matthew.wall@imperial.ac.uk

C. R. Madan (ed.), *Academia and the World Beyond*,
https://doi.org/10.1007/978-3-030-82606-2_3

Chris: Can you introduce yourself and tell me a bit about your current position?

Matt: I'm Matt Wall. My main employer is currently a company called Invicro, which is an international contract research organization (CRO), meaning we do research projects for other commercial entities, mostly companies in the pharmaceutical and biotechnology industries. We have a few sites in the United States, one in Europe (London), and a small office in Japan. I work at the London site, where we also do a lot of academic work, mostly with people from the main London universities (Imperial College London, University College London, and King's College London). Because of my involvement in various bits of academic work, I also have an honorary senior lecturer appointment in the faculty of medicine at Imperial, and I am an honorary research associate at UCL as well. My area of expertise is neuroimaging using functional magnetic resonance imaging (fMRI), so most of my job is based around that technique, but within that, it can be pretty varied, and at any one time I'm juggling multiple ongoing research projects—some commercial, some academic—which are at various stages of development, data acquisition, analysis, and being written up. These range from academic neuro-endocrinology projects, testing out new drugs for commercial companies, more technical projects focused on acquisition or analysis methods, supervising master's or PhD student projects, and collaborating on studies looking at the brain effects of cannabis, MDMA, psilocybin, or other drugs.

What was the focus of your PhD?

I started my PhD in 1999 in the Department of Psychology, University of Cambridge. Because my supervisors moved during my PhD, I actually finished it at the Department of Psychology, University of Southampton, though I stayed a registered student at Cambridge, and officially got my PhD from there. My PhD work was on pretty straightforward cognitive psychology, and I used reaction time tasks to investigate the possible role of implicit learning in anxiety and threat perception. I'd done an experiment on implicit (unconscious) learning for my undergraduate research project and was really keen to try and apply that to different sets of stimuli for my PhD work. That was perhaps my first major mistake with my career; my supervisors (very sensibly) suggested an alternative project that probably would have been somewhat easier and more achievable, but I was very keen to focus on implicit learning and basically ignored their advice. It turned out to be a very difficult topic for lots of reasons, and I got a grand total of zero papers published from my PhD work.

As you were finishing your PhD, what were you thinking about your career plans?

My PhD work was not stunningly successful. I managed to write it up into a fairly coherent thesis, but I didn't have much else to show for those years of work, so I felt I wasn't a strong applicant for research positions. Toward the end of my PhD, I was considering all kinds of options. My girlfriend at the time was a medical student, so for a little while, I was even looking into graduate courses in clinical medicine. At that time in the early 2000s, there was a little bit of a revolution going on in cognitive neuroscience. The early work with fMRI in the mid-to-late 1990s had generated a lot of excitement, and a lot of departments were interested in it and some were even buying equipment, but there were still relatively few people who knew how to work with those techniques. I was as excited by this new technique as everyone else and realized that it was going to be an important method, so I decided I would try and get a research job using fMRI.

I got interviews for a few post-doctoral positions—in Cambridge and King's College—but with no success. Then I applied for a Wellcome-funded postdoc position at Royal Holloway and was lucky enough to get an offer there. The psychology department at Royal Holloway had just bought a new MRI scanner and needed staff to use it.

Can you tell us a bit about what day-to-day life is like in your current position?

It's pretty varied. I currently have one research assistant and PhD student who I work with very closely, and between us, we're responsible for all the fMRI projects that are running at our facility, so we have to do everything. On any given day, we might be writing stimulus programs for experiments, running pilot scans or doing other technical testing of the scanners or stimulus equipment, assisting the radiographers with data collection, or running data analyses. I may also be talking to customers and preparing proposals for potential studies, or attending regular meetings. There are monthly or fortnightly meetings that focus on planning/forecasting, potential new projects, and various business units like the MRI group, or the analysis group. Plus there are a number of other PhD students that I help co-supervise and quite an extended network of academics that I work with on various projects, some of which I have regular or semi-regular meetings with. At some points during the year, I also supervise Imperial MSc students, so I try and spend as much time as I can with them, particularly at the beginning of their projects where they need some help and teaching. Occasionally, I might go somewhere else in London to meet someone else or give a talk or something, but usually everything's pretty local.

I try to have one day a week working at home where I'm supposed to focus on writing papers. I usually manage to stay at home, but very often I end up using that time to catch up with all the admin tasks and emails that have accumulated that week, so it's only semi-successful. At least a few times a week, I tend to crack open my laptop at about 9 pm (after my kid has gone to bed) and do a couple of hours of catching-up work in the evening.

What do you like most about your work?

The variety of things I work on and the pretty large network of people I end up working with. The fact that, unlike academic scientists, I have very little pressure on me to write grants or publish in high-impact journals. I take on as many students as I want, and I never have to sit down and mark a stack of essays or exams. I like doing the research myself, but as I get older, I've found I really enjoy a mentoring role too; my first PhD student did some fantastic work throughout her degree that we're both really proud of, completed successfully, and has moved on to a great post-doctoral position at Oxford, and that's been incredibly rewarding for me.

And what do you like least about your work?

While I don't have the pressure to write grants, I am under constant pressure to bring in more commercial work from external companies, and that can be difficult. Then, when we do get a commercial project from, say, a pharmaceutical company, there is typically a very large team of people working on it, sometimes across multiple companies/organizations at different sites around the world, and everything proceeds extremely slowly and with a huge amount of bureaucracy involved. Academic teams tend to work like agile start-ups: they don't have a lot of money, but they tend to move fast and get things done. Working with large companies is the opposite, and sometimes, it can be incredibly frustrating. They also tend to be quite conservative and risk-averse in lots of ways, but particularly when it comes to their approach to research. Academics want to try exciting new things because they want to be at the cutting edge, but commercial customers often want to do an exact replication of a task that's been done hundreds of times before because it works and they're likely to get good results.

How do you think having a PhD has helped you succeed in your current position?

There were two things my supervisors were insistent I learned during my PhD that I'm grateful for all the time. Firstly, they made sure that I was able to program my own experiments. I learned to code in C and also in this horrible old DOS system called MEL (Micro Experimental Laboratory). I never use C or MEL anymore of course, but the principles led me on to learning MATLAB, Python, bash scripting, and various other things, which has been such an enormous benefit throughout my working life. Secondly, they taught me about really good, efficient, experimental design, which is also knowledge that I've relied on a lot in my career and have tried to pass on to others.

Now, I guess I could have acquired those skills and that knowledge without doing a PhD, but it seems much less likely. I've known people who have worked in similar positions to me who didn't have a PhD, but they're rare. Most science-led businesses want scientists for employees, and that means people who've had scientific training, which usually means a PhD.

If someone currently finishing their PhD was considering a position similar to yours, how might they decide if it would be a good fit?

It's tricky trying to decide on a commercial/industry career. I actually think my current position is fairly unusual, in that it's a commercial company, but I still get to work on a lot of academic projects. For me, at some point, I realized that what I really enjoyed was the process of doing research: figuring out design issues, squashing programming bugs, trying new ways of analyzing data, that kind of thing. My current position has a certain amount of corporate annoyances associated with it—nothing's perfect, after all. But it lets me work with really smart, engaged people, and I get to do all the fun parts of research without having to worry too much about where the comes from. That means I give up some autonomy and end up working on other people's projects somewhat more than my own, but I like the variety.

A lot of people like academia because they feel it gives them an opportunity to work on a topic that they deeply care about. Do you think this is also true in your current position?

Somewhat. I certainly get to work on a lot of different things that I find very interesting, which is great of course, but it means that I've never really developed a sustained line of research on any particular topic. My publication list is really varied and starts with lots of low-level vision research from my postdoc years and then meanders into psychopharmacology and pharmacological fMRI, with some neuroendocrinology and some methodological papers, plus a few other random things scattered throughout. Through taking up whatever seemed to be the best job opportunities that were available at the time, I've ended up being more of a specialist in the methods I use, than on any particular research topic, and I'm basically fine with that. I find most topics in psychology and neuroscience pretty fascinating, once you get into them enough. This is something I often find myself telling younger colleagues, particularly when they're applying for PhDs: Don't get too hung up on any particular topic. Some people have a burning desire to work on, for instance, psychosis and will only apply for positions in that area, which I think is a mistake. PhD and post-doctoral positions are incredibly competitive these days, and you should apply for everything you can find. You may end up working in an area that perhaps you hadn't considered before, but that often gets you some different experience and perspectives. Maybe you get back to your favorite research topics later in your career, or maybe you carry on working in the new area because you find you enjoy it. It's good to have some kind of plan, but often you need to be somewhat flexible and realistic about your options as well.

If someone was interested in pursuing a similar career path, what would you suggest they do to better prepare themselves?

The number one thing: learn some useful skills. Most of the opportunities I've had in my career have come about because I've had skills that have been useful to people. I got my post-doctoral position because I could do a bit of programming, and a lot of my collaborations have come about because I've offered to program tasks or run analyses for people. Learn how to program: it doesn't really matter which language, but try and learn a couple! Learn how to use Unix/Linux systems and some bash scripting; learn how to design good experiments; make sure your statistics knowledge is solid and up-to-date; learn how to make great-looking figures with image-editing software; learn about pre-registration and the open science movement. Anything like that. Really solid technical skills are the most important things I look for when hiring people. You may know everything about the literature on a particular topic, but that knowledge is specific and of limited usefulness if you end up working in a different area. Technical skills are always useful, no matter what you're actually working on.

Based on your journey, what advice or suggestions do you want to pass on to someone who's currently finishing their PhD?

With due consideration given to blind luck and both survivorship and hindsight bias, I think if my career demonstrates anything, it's the virtue of being flexible about your research. I've never had much of a plan for my career and have generally taken jobs that were convenient at the time, with relatively little regard for the exact research topic. Possibly that suits me more than others, as I found myself interested more in the methods than any particular research topic, but I still think it's a valid approach to a career; science needs expert methodologists as well as specialists in particular topics after all. Attention to learning useful research skills early on in your career can help in following a path like this, if you think it might be for you too.

Thank you so much for sharing insights into your journey with us; it is greatly appreciated!

"Align Your Mission and Vision with Your Next Steps After Your PhD"

Maira Quintanilha

Maira Quintanilha

Abstract Here, we interviewed Maira Quintanilha about her path from PhD to founding her own business—helping academics and researchers with their qualitative research projects. Having initially planned to be an academic, Maira found a new path after realizing she could build a business to teach qualitative methods online, a career path that would offer more flexibility and allow her to prioritize her family. Maira was drawn to academia to learn and inspire learning in others and now does that through her business. Maira shares many insights she learned along the way, such as finding your "zone of genius" and mentors, considering what you are passionate about, and investing in your own skill development.

Contents

M. Quintanilha (✉)
Quali Q Inc., Edmonton, AB, Canada
e-mail: maira@qualiq.ca

C. R. Madan (ed.), *Academia and the World Beyond*,
https://doi.org/10.1007/978-3-030-82606-2_4

Chris: Can you introduce yourself and tell me a bit about your current position?

Maira: My name is Maira Quintanilha. I was born and raised in Brazil and moved to Edmonton, Canada, in 2009 (where I currently live with my husband and two children). I'm the founder and owner of Quali Q Inc. I help academics, researchers, and clinicians to design, implement, and write impactful qualitative research projects in health sciences and interdisciplinary fields related to health.

What was the focus of your PhD?

Officially, my PhD was in Nutrition and Metabolism at the University of Alberta, Canada (Department of Agricultural, Food and Nutritional Sciences). Yet, I have a hard time seeing myself as an expert in metabolism because my PhD work was in maternal health during pregnancy and postpartum with a very strong focus on social determinants of health and qualitative methods of inquiry. I learned to navigate the intersect of social and health sciences during my PhD, often standing up for my methods in a primarily quantitative department while not being well versed in social theories.

As you were finishing your PhD, what were you thinking about your career plans?

Oh, this is going to be a long one. I basically spent three years "finishing" my PhD, and in those three years, a lot changed. In July 2017, during my third and what could have been the last year of my PhD, I got pregnant with my second child. I had a Vanier Graduate Scholarship and had to make a choice: defend before the baby was born or use my second maternity leave privileges and defend upon my return. I chose the second option, I was working with one of my supervisors, Dr. Maria Mayan, in the second edition of her book *The Essentials of Qualitative Inquiry* and loving it. I had decided to build a strong qualitative research portfolio and eventually pursue an academic position, so using my Vanier to build that portfolio made sense. In July 2018, when my daughter was three months old, I signed a contract with the Faculty of Extension to teach a graduate level course in qualitative methods in January 2019 and started getting ready to return to work in November 2018. I had a solid plan—teach the course, work on my thesis, defend, and seek either a postdoc or research coordinator position at the University of Alberta (moving somewhere for a postdoc wasn't an option as my husband owns an engineering company here).

Then, everything changed in September 2018 when Elizabeth (my then 5-month-old daughter) was diagnosed with cancer. I lost my ground, my desire to work, to be a successful academic and for a while to finish my PhD. I pushed the thought aside many times, but it insisted on showing up frequently: "if I lose my daughter, what's the point in having a PhD?" As much as I suffered during that time and until the end of her treatment (July 2019), I learned to ground myself in the present, to use mindfulness strategies and techniques to deal with PTSD, to tap (another mindfulness technique I learned from the psychologist from Kids with Cancer), to breath, and to thrive amid the worst few months of my life. It was in those months when it was mostly her and I at home during the day (my son was going to daycare) that I started allowing myself to think about my future. I'm an optimistic, a maximizer, but I knew academia was not the best place for a mother who had a child who could require months of care in the future and needed the ability to drop it all and be at the hospital. Besides, I had decided I wanted a bigger family (which meant more kids and time off).

One day, I was browsing the Minimalist Baker website and came across their online course on how to take better photographs of food. I felt like I had won the lottery! I could build a business to teach qualitative methods online while being outside of the demands and expectations of an academic institution. Plus, I would have the flexibility I needed to prioritize my family. Little did I know how much work it would take to build Quali Q and what it is now (and what I want it to be two years from now). But I had a vision—I was dreaming again and it was all I needed to not give up on my career and my PhD. I started taking online business and marketing courses while on medical leave with my daughter, and when I returned to graduate school in September 2019, I had already established Quali Q Inc.

I defended in January 2020, and once my thesis was submitted and approved, I turned my attention to my business and it's what I do now. I never imagined that I would be an entrepreneur or a consultant, but I would say I have been "becoming" one ever since I decided that would be my post-PhD plan. It felt so good to be able to tell people I had a post-PhD plan. I felt relieved, privileged, and empowered to have that answer because let's be honest, PhD candidates, especially those who are close to defending, often dread the question "what's next for you?"

Thank you so much for sharing your personal story of how your plans changed. Can you tell us a bit about what day-to-day life is like in your current position?

I work approximately 30 h/week, and a big part of my work involves writing. With the decision to build an online business comes a commitment to creating consistent content that can serve the Quali Q audience. My weekdays involve meeting with current mentees or prospective clients for the one-on-one mentorship program, writing blog posts and weekly newsletters, and reviewing any work my virtual assistant does. To write blog posts and create the qualitative courses, I have to study and keep up with the literature so that I can learn new ways to teach concepts to my students and audience.

What do you like most about your work?

Two things drew me to academia: learning and inspiring learning among others. I get to do that now through Quali Q. I'm constantly learning about qualitative methods, marketing, habits, and behaviors while also supporting other people's learning and growth. It's what I like the most about my current role.

I believe all graduate students should understand what drew them to academia and what skills they have learned that could be applied to other work areas. It's important to understand their unique genius and the skills they used during their PhD that are within that zone of genius. The zone of genius is a reference to *The Big Leap* by Gay Hendricks. In the book, Hendricks talks about how there are tasks we're good and great at, and yet they might not be activities that we'd do every day and thrive in our careers. Finding our zone of genius requires looking within ourselves and critically thinking about what motivates us, what makes us feel energized about doing as part of our work and lives, and how we see ourselves contributing to society. If our work reflects our zone of genius, we're more likely to thrive in our roles and reach our full potential.

And what do you like least about your work?

Being a business owner comes with many responsibilities I don't like but need to do, including bookkeeping, understanding taxes in Canada, and, internationally, having difficult conversations with people about money. I'm not in this to be rich, but I need to make a living and sometimes those conversations need to happen even if I don't feel like having them.

How do you think having a PhD has helped you succeed in your current position?

This is such an interesting question. I don't think I would have been able to do consulting without my PhD. I had a rich PhD experience that gave me both the know-how and confidence to start Quali Q and carry the responsibilities of my current position. I also think that having had success in acquiring funding and publishing as a PhD student boosted my confidence to support and inspire others. I do believe it's possible to build a successful career as a qualitative researcher, and this is something that came with my PhD experience. Another thing that I took advantage of during my PhD that has helped me succeed was the mentorship program offered to graduate students at the University of Alberta. I had one mentor in 2017 and another one in 2019, and they were both instrumental in decisions I made that led me to where I am now.

Can you tell us a bit more about what sort of questions you asked your mentors for advice on?

My first mentor had a PhD and was working outside of academia, so my questions to him were mostly about how he had transitioned from academia to government. It was so interesting because we sat at a coffee shop with my highly academic CV on the table and a few job postings available at that time, and started thinking about how my academic achievements would enable me to apply for those jobs. That exercise was powerful because it made me realize that I would not be transferring publications and awards to any job application outside of academia. However, I could transfer the skills that helped to publish the work and win the award, such as writing, facilitation, and communication skills. One of the skills that we identified as being a potentially useful one (for what I wanted) was qualitative research expertise. And guess what? I loved that! When we finished that mentoring relationship, I had actually decided to pursue an academic career post PhD, which changed again after my daughter's diagnosis.

My second mentor was a consultant, and when we met, I had already started Quali Q Inc. Our conversations were more about implementing systems in the business that would make it more effective and efficient. She was also a mother of three children, so it was very inspiring to learn how she integrated—not balanced—her personal and professional lives.

If someone currently finishing their PhD was considering a position similar to yours, how might they decide if it would be a good fit?

I think they would need to really think about their zone of genius and whether what I do in my day-to-day work aligns with their skill set and vision. When it comes to being a consultant or entrepreneur, I do believe that there are some essential traits one must possess. First, a sense of adventure and comfort with a certain level of risk. You never know if your first business idea will work, and it takes a couple of years to start seeing results and financial rewards. Being resourceful and persistent also helps a lot. I've learned so much about so many things (from website coding and social media platforms to marketing and copywriting), and as proud as I'm to say that, sometimes, it does feel overwhelming.

If someone was interested in pursuing a similar career path, what would you suggest they do to better prepare themselves?

I started thinking about my business model and services during my PhD and would recommend PhD students to do that. Going back to the idea of the zone of genius, I think they need to think about the skills they built during their graduate studies that were strong (within their zone of genius) and could be sold in the form of a product or a service. For those with product ideas who need capital, I would look for grants and local agencies that might offer support to small businesses. In Alberta, we have the Alberta Women Entrepreneurs (AWE) and Edmonton Start Up, for example. For me, I started taking online courses about how to create and sell online courses. I also began to read business and marketing books as well as listen to podcasts from marketing experts and successful online entrepreneurs. Interestingly (and not surprising to me), the thing that made the biggest difference in my current position was finding a mentor who does something similar to me with a different group and product. Learning from her has been inspiring and instrumental. In other words, seek mentorship and embrace the fact you're a lifelong learner.

A lot of people like academia because they feel it gives them an opportunity to work on a topic that they deeply care about. Do you think this is also true in your current position?

Absolutely and, dare I say, even more than I would inside academia. I get to choose the projects I say yes to without the pressure for funding or political forces. In the beginning, I said yes to things that now I no longer do, and it's part of growing and becoming more established. I believe that the reason I get to do what I deeply care about is because I looked deep inside me when I decided what I cared about and wanted to continue doing post PhD. I'm a dietitian by trade; however, I knew that nutrition wasn't the topic and skill set I wanted to develop and use in my next position. Could I do well in the world of dietetics? Oh yes! Is teaching nutrition or offering dietetic services in my zone of genius? No, and I recognized that. I recognized that what sparked passion, curiosity, and enthusiasm was qualitative research methods; thus, the skills I'm using in my current position reflect that. I think stepping outside of academia means discovering new ways (and perhaps traveling the least travelled paths) to work on things and topics one deeply cares about.

Another reason many like academia is that they feel it provides them with more freedom than they think they would get in other positions. How much freedom do you feel you have to work on what you think is interesting?

I absolutely agree with this. I think academia offers freedom and flexibility … even more once you become tenured and have consistent funding. It was always one of the things that I loved and appreciated about academia. It's also the main reason why I decided to start my own company rather than seeking a position in government or industry. In the beginning, the financial pressure (in other words, the need to not sink money into Quali Q) diminished my freedom to choose what to say no to, but not my work flexibility. Almost two years into business, I've had more freedom to do the things that I enjoy. My expectation is that as I become more established, I am more able to delegate responsibilities that I don't spark joy in me and outside my zone of genius … the best example I can give is bookkeeping.

Based on your journey, what advice or suggestions do you want to pass on to someone who's currently finishing their PhD?

Many people will tell you to look for people whose career you admire and try to emulate it. However, I would say there is something even more important that should happen before that. Look within yourself. What do you want your life to look like? What are your passions? What are your (current) goals and priorities? What matters most to you? What is your personal mission? These are not superficial questions; they require thinking, reflecting, talking to others…. If you don't know yourself and what you want well enough, it's easy to get lost in the course of a career transition and end up working outside of your zone of genius. You can probably do it, but you might not thrive (and let alone have fun along the way).

Then, align your mission and vision with your next steps after your PhD. When I had my first mentor, before my daughter was born, I had decided to pursue an academic career, and my plan was to seek a postdoc or a position as a research coordinator after my PhD. When my plans changed, I shifted to defining Quali Q Inc. services and the clientele and focused on learning how to start and run a business. However, those practical steps came after I decided what I wanted and what I was trying to build.

Is there anything else you'd like to tell someone reading this interview?

Invest in building your leadership skills throughout your graduate program. You might not see yourself as a leader, but within a certain circle of influence, you surely are. Leadership skills will build your resilience, improve your ability to communicate, boost your confidence and creativity. No matter your personality type, you can learn to be a great leader of people (think of books, podcasts, mentoring programs, volunteer opportunities, etc.), and that will be the best skill you'll hone in during your PhD. Whether you choose to stay or leave academia, strong leadership skills will open doors for you, and the right opportunity through one of those doors is all you need to find your own path post PhD. It's not always easy. Yet, how many great things in life are easy or come to us easily? Yes, not many. So keep at it while being true to yourself.

Thank you so much for sharing your experiences and perspectives with us, Maira!

"I Might Have More of a Knack for Science Communication Than for Doing Actual Science"

Jens Foell

Abstract In our interview with Jens Foell, we discuss his prior research work in neuropsychology and neuroimaging and recent transition to a full-time science communication position. For several years, Jens has been involved in a large-scale "rotating curator" science communication project called Real Scientists. More recently, Jens led the launch of the German language spin-off. Jens shares his insights in how to be a science communicator and how this has become a full-time job for him. Jens also discusses how labs can differ in their culture—such as their working environment and expectations for working hours.

Jens Foell

Contents

J. Foell (✉)
Department of Psychology, Florida State University, Tallahassee, FL, USA

maiLab, Heidelberg, Germany
e-mail: foell@psy.fsu.edu

C. R. Madan (ed.), *Academia and the World Beyond*,
https://doi.org/10.1007/978-3-030-82606-2_5

Chris: Can you introduce yourself and tell me a bit about your current position?

Jens: Up until very recently, I had a pretty standard research position: I was an Associate in Research at Florida State University. I was with the same department for almost eight years and have been involved with different interesting projects over this time, including helping to build up a new MRI imaging facility and investigating phenomena such as aggression/impulsivity, fearlessness, cognitive flexibility, depression, and pain. Before that, I worked on my PhD back in Germany. The transfer from there to Florida was facilitated by working on a paper that had both my PhD mentor and my then-future boss as co-authors. Apart from the change of scenery, it was also fascinating to see the differences in lab culture: at least on paper, my line of research is the same all over the world, so it's quite easy to move countries and start in a new lab. But of course every lab has different expectations and different ways to approach the same thing.

However, just over the last few months, my life and career have taken a big turn. The start of this was that during my time in Florida, I have gotten more and more into science communication and have written articles, set up a Twitter platform for scientists, given public talks, and so on. Importantly, this was always at the level of a hobby; something I did for fun in the evenings and on weekends. But I noticed that I might have more of a knack for science communication than for doing actual science. It felt to me a bit like the difference between building a car and driving it: doing research shows you the nitty-gritty of it, why things work the way they do, similar to designing and constructing an engine. You have to know your way around, and you might get dirty. On the other hand, if you're driving the resulting car, you don't necessarily have to know how the engine works. It requires an almost entirely different set of skills and understanding, but doing it well can still be hard. And you can always find yourself in a situation where knowing a lot about how the engine works ends up being quite useful.

The big turning point arrived when I was invited to Germany to present a German-language science communication project that I set up: Real Scientists DE is a spin-off of the popular Real Scientists Twitter account, on which every week a new scientist talks about their methods, their life, and their research findings. What I did was to help expand the concept by setting up the same thing for

German-language researchers. This entailed creating the actual account, reaching out to scientists and scheduling their curation weeks, and trying to promote the account in the German-language research and science communication communities. The main motivation was to provide a service for an audience that might not be fluent enough in English to follow science content. In some ways, the German-language landscape for science communication sometimes seems less developed or less extensive than the English-language market, so I figured there might be an eager audience for it. I was very happy with the result: apart from a steadily growing audience, there was a lot of fascinating science that was being shared, and everybody seemed to have fun with the account.

That's also what I told at the meeting in Germany. It had been organized by a science communication institute and brought some of the biggest names in German science communication together. Among them was Dr. Mai Thi Nguyen-Kim, an MIT-trained German chemist who started out as a science YouTuber and presenter but very quickly grew to be the most important science communicator in the country. Her accolades include a best-selling book, virtually all major media and journalism awards, the highest medal of merit that can be awarded by the German President, and a YouTube channel with more than a million subscribers. At the time, I already knew a member of her team (again, the German science communication landscape is smaller than its English counterpart), and we stayed in touch after the meeting. A while later, she looked to expand her team, and we started talking about a position for me.

This position has brought me and my family back to Germany just a few short weeks ago, and I am still learning the ropes. The transition back after almost a decade abroad can be a bit disorienting, as it turns out, and developing content for a million-subscriber platform is just as dizzying. Being already familiar with the team certainly helps, and so does working for a tremendously talented and yet very down-to-earth boss. My actual job consists of everything from keeping an eye on current trends in order to identify interesting topics, to researching familiar and unfamiliar scientific topics, to reaching out to researchers for background conversations or interviews, to coming up with ways to frame scientific content so that it is accessible and entertaining. The skills I've developed as a researcher are surprisingly helpful for this, especially when examining the methodological quality of a published study and when chatting with researchers from various fields.

That was a great introduction. I have lots to ask you about! Let's start from the "beginning." What was the focus of your PhD?

I had the unique chance to do PhD research on a fascinating topic in the lab of one of the world's leading experts on it: studying phantom limb pain under Dr. Herta Flor in Mannheim, Germany. I had developed a strong interest in neuropsychology during my undergraduate studies, and I was particularly interested in topics that had the prospect of tangible results, such as something that might be able to actually help people in the foreseeable future. The core of my PhD thesis was the following: by the time I started, we knew that the intensity of phantom limb pain is correlated with the amount of cortical reorganization after amputation. This reorganization can be measured using neuroimaging, and its relationship with pain is something that had been discovered by Dr. Flor in the mid-1990s, using the comparatively crude fMRI brain scan technology that was available at the time (and the finding has since been replicated and expanded). In other words, while the best and most precise measure of pain is still asking someone how much pain they're in, at least for phantom limb pain, we seem to have a reliable physiological correlate at our disposal. The direction of this correlation, i.e., whether changes in the brain cause phantom limb pain, or whether persisting phantom limb pain changes the organization of the cortex, is still being debated.

Also, shortly before I started my PhD work, it became known that a simple and cheap procedure might be an effective treatment for phantom limb pain. What I mean by that is the so-called mirror box therapy, in which a patient after amputation is presented with the visual illusion of the lost limb being back. After having been anecdotally reported to work for around a decade, the early 2000s saw the first controlled evidence for its effectiveness. With these data available, as well as a state-of-the-art fMRI setup, I wanted to investigate whether successful mirror box treatment would be able to reverse some of the cortical reorganization that has been associated with phantom limb pain. Spoiler alert: that's indeed the case. Post hoc findings included a lack of correlation between mirror treatment effectiveness and time since amputation (i.e., how long a patient had already been living with phantom limb pain did not predict treatment success in either direction) and some speculation on whether properties of the phantom limb itself might be impeding treatment success in some.

To my delightful surprise, the resulting paper would later win me a prestigious German pain research award, which might very well remain the pinnacle of my research career (at that time, I was already living in Florida, and the fact that I wasn't able to travel back for the award ceremony still bugs me considerably). During those studies, I got involved with others that were closer to the patient experience, like helping develop an augmented-reality-based mirror box system and evaluating new prosthetic devices.

How did you decide to move to the USA? Was the cross-lab collaborative project first, or did that emerge after you moved?

Right after my PhD thesis, I was at a crossroads both regarding research topic and regarding where my wife and I would live and possibly start a family. We had already lived in the USA before, and doing that again seemed like a fun thing to do. The other big neuropsych research topic that had always interested me was psychopathy, or aggressive/delinquent behavior in general. It touches on several important societal topics (most prominently the question of whether our brains are "hardwired" in some way to go against society). As coincidence would have it, I became involved in a collaborative psychopathy project between Dr. Flor and Dr. Christopher Patrick in Tallahassee, Florida. At the same time, Dr. Patrick was looking for ways to include more neuroimaging in his lab work. This situation developed into me moving to Florida to help build up an fMRI component for the lab while also conducting non-fMRI research in the lab, such as an investigation into cognitive flexibility when confronted with similar tasks requiring different sets of rules.

While Florida is a very different place from Germany, there has always been a healthy exchange between the two regions, and overall we had an easy time fitting in. However, a few years after transferring there, the country fell into a period of political turmoil. By that time, we had already started a family in Florida but also kept up strong ties to the home country. The position also allowed me to branch out into interesting side topics, including working with the defense team of a prominent criminal case—specifically: answering the question whether the defendant's clinical MRI scans could somehow be used as part of their psychopathy assessment. At the same time, I tried to hone my skills as a communicator, which was facilitated by such things as a locally organized TEDx event and by being in the same time zone as many interesting and friendly scientists who I started to interact with on Twitter.

As you were finishing your PhD, what were you thinking about your career plans?

The most obvious thing to do at that stage was to start a postdoc in the field of pain research, to further carve out a niche for myself and work toward a professorship. But as may or may not already be apparent based on what I've stated so far, I've never had a clear career plan cut out. I've always followed the topics and methods that interested me most out of those that were available, based on a conviction that those will also be the ones that I would be best at. So far this approach has worked, although it's hard to tell (a) whether this has been mostly due to sheer luck and an overestimation of my abilities by the people making hiring decisions and (b) whether there will come a day on which I will regret this zigzag course and wish I would have stuck with one topic and area. But at the very least, I can demonstrate that a clear career plan is not a necessary condition for doing research and science communication in interesting labs and locations.

One factor that might be important to understand my decisions is that I did not grow up in an academic context and so its career paths have always felt somewhat alien to me. So even when I was working toward building an academic portfolio for myself, the prospect of being a tenured professor somewhere never seemed as desirable to me as it might be the case for others.

Based on your journey, what is some advice or suggestions you would want to pass on to someone who's currently finishing their PhD?

One thing that I have definitely learned is to try and find the environment (including topic, methodology, and workplace conventions) that fits you best. An example of what that might mean in a research context is if you're more or less okay with fuzzy data or incomplete interpretations, or if these things make you uncomfortable and you'd rather have strictly defined problem spaces and solutions. Either of these things is available in today's scientific efforts, and I think it's important to find the landscape that fits you. And by "fit," I don't necessarily mean the one that is identical with your own way of doing things—you might actually seek out the other way so that you'll find work that challenges you on a day-to-day basis. In summary, this mainly means that you shouldn't underestimate your own gut feeling, even in a head-centered enterprise such as scientific research.

For science communication, this extends into the medium that you will be using. Maybe you're someone who is good at writing long-form articles or bite-sized commentary. Maybe you have a good voice for podcasts. Or maybe you're a happy extrovert and most comfortable when you're jumping on a stage and giving people a funny impromptu rundown about your scientific field. The important thing to remember is: there is an audience for each of these things. That means that you might find it useful to try one or the other and see what works best for your style. Knowing how best to express your own voice, both in terms of content and medium, is an incredibly powerful skill. Or to phrase it differently: if you're funny, be funny. If you're cold and analytical, be that. Trust that someone out there will find your way of framing things the most accessible one. Challenge yourself to find something new, but don't contort yourself in order to fit into a niche that's not yours.

One notable difference between working in research and working in science communication is how often you will change your topic. In academia, it's common to work on the same tiny aspect of your subfield for a decade or more. In my current sci comm job, I might look into a topic for 2 to 4 weeks before moving on. Again, it's not like one of these things is inherently better or worse; it's more about balancing your own preferences for stability or flexibility and finding or choosing what suits you best.

Can you tell us a bit more about Real Scientists and then the DE spin-off?

The Real Scientists project was initiated by scientist and science communicator Upulie Divisekera in 2013. Its primary component is a Twitter account that changes hands every week. Even if you're entirely unfamiliar with Twitter, it's likely that you have encountered this part of the platform before: because another example for this type of account is that for the US President, which is assigned to whoever currently holds that position. And just like the account for the President, the Real Scientists account is a very direct way into the mind of whoever is currently at the helm. The concept is known as *rocur* (short for rotating curated) account. One big advantage of a rocur like Real Scientists is that as a reader, you'll get to know someone new every week. So if you find yourself disinterested in the current curator or their science, you can just wait for the next one. And over the weeks and months, you'll get an overview of the entire spectrum of scientific fields and personalities. And it's free! It's no surprise that after the initial success of Real Scientists, other science-themed rocurs have started to appear. But Real Scientists remains the oldest and the one with the broadest appeal, as it doesn't tie down the scientific area that is being displayed.

The project fascinated me right away when I set up a personal Twitter account for myself in 2014. What makes Twitter unique as a platform for science communication is how direct and unfiltered it is—for better or worse. It's a lot like chatting with people at a scientific conference. You might not understand everything that's being said (and if you do, you certainly won't agree with all of it), but you'll gain the most current possible insight into what bothers or excites the experts at any given moment. Also, as you know if you've ever been to a science conference, the social aspect is not to be underestimated; and so it can also be fun and useful to chat with scientists about personal interests or leisure activities.

The moderators behind Real Scientists were kind enough to have me on as a curator for a week in 2015. It was a quite exhilarating experience; at the time, the account had a bit over 20,000 followers who were eagerly asking questions about my work and discussing my field in general. Shortly after that, I was offered the position of a moderator myself, which meant that I had the opportunity to learn how the account works behind the scenes. At the time, there was already the idea of branching out into other languages, and I was excited to spearhead the first spin-off. Given my native language and the lack of anything like Real Scientists for a German-language audience, it felt natural to launch Real Scientists DE, which I did in early 2017.

The most important thing to know in this context is that providing the infrastructure for the project is not difficult at all. Instead, the biggest potential bottleneck is finding curators who provide content that is interesting and regular enough for the audience to keep tuning in, which is the only way to grow your project's user base. The other important thing to know is that, while it is idle to only focus on the number of your followers, a larger user base will help you to attract new curators, which in turn grows your base—in essence, you want to end up with a self-perpetuating system. But that means that the fate of your project is mostly not in your hands but in those of the curators that you invite (or that you schedule after they reach out to you). And luckily, Real Scientists DE has had a constant level of wonderful curators and has been able to grow its base at a steady pace (to currently about 11,000 followers, about 3 ½ years after launch).

Is there anything particularly interesting that you learned about through someone else's curation of the Real Scientists DE account?

In terms of science facts, I learn new and interesting things basically every week. But what has been even more striking was learning about the researchers' own motivations and concerns about science communication. After talking to and scheduling around 200 curators over the last few years, it is easy to recognize patterns in the questions that they have before taking over the account. Many of them worry about not tweeting enough to keep people entertained and excited, when in fact this doesn't seem to be a problem for the audience at all: if there's a lull, people just tune in the next day. And if they're not interested in your field of research, they'll be back next week. The thing that might actually cause trouble is likely something that you won't see coming, like a statement about your personal beliefs about science that parts of the audience might be opposed to. It's rare to see a debate become uncomfortable or nasty on the account, but if that happens, it's likely about a current hot-button issue. Overall it has been most interesting for me to see what catches curators by surprise and how they might deal with it. Luckily, neither the English-language account nor the DE account has ever seen a big meltdown.

The other very interesting thing is to see where curators go after their week. Often they're people that I still stay in touch with on Twitter after, and for many of them, Real Scientists DE is a first gateway into sci comm. As a result, I have seen several of them switch to full-time science communication later, and whenever possible, I have tried to have them reprise their role as a curator after their transition, which provides an interesting perspective.

Can you tell us more about how you realized you would rather pursue a science communication career rather than staying in a research position?

It started early but still took a long time. Right when I started my undergraduate studies, I noticed that it was a lot of fun to talk to others about the psychology and neuroscience studies that I learned about in class. And I quickly realized that the questions and comments that came back from my friends helped me sort through my own classes as well. In other words, in order to explain something well, you need to understand it well yourself. One fascinating example for this is statistics: in undergrad research, it's most important to be able to run the analyses you need to run and to clean and organize your data in a useful way. But when talking to non-specialists, they will very soon ask you *why* you are running the analysis, or if there are other ways to find out the same thing. To answer that, you'll have to take a step back and see the whole picture, which is often very helpful. At the time, I didn't think about a career in science communication, but I felt myself gravitate toward situations that allowed me to talk about these things.

My first moment of testing the waters came in 2009, during the first half of my dissertation, when I offered a science class in an evening school. For this, I combined information from psychology and basic science to create a course on critical thinking and hypothesis testing. While there was never much interest in the class, the people who did take it always gave me positive and useful feedback and often asked for additional information afterward. This experience taught me not only that I have an interest in doing science communication but also that it's important to have an effective platform. Over the few years that I offered the class, attendance ranged from something like 8 to 12 people, which didn't feel like a giant audience (although I was grateful for everyone who was there). This is something that changed from 2014 on, when I started being on Twitter. I was lucky to be accepted into what people often call "science Twitter," which is a circle of current and former scientists, as well as science enthusiasts, that regularly interact with one other on the platform. Becoming part of this group allowed me to build a relatively large audience fairly quickly. This provided me with a new playground to test science communication ideas: like what does a science fun fact have to look like so that people will read or share it? What hook do you need at the beginning of a longer thread to make sure the audience will read the whole thing? Mind you, some of these questions I still can't answer even after more than half a decade on Twitter. But over time, I realized that thinking about these things was much more fascinating to me than, say, writing a grant proposal, which is a surprisingly large part of having a research career.

What is day-to-day life like in your current position?

We have a lot of virtual meetings and chats within the team to develop new ideas for topics. Usually there are more topics available than needed, so that some might be put on a back burner or might be dropped entirely. Once a topic has been selected, we pick the best person to take the lead on it. Since the team members all have different backgrounds, this is often an easy question. After that, the part that I spend most time on by far is researching the topic at hand. One thing that has made this particular channel popular is that the conveyed information always represents the best and most current science. Making sure that this is the case might entail several extensive literature searches and/or reaching out to experts for chats and interviews. My main job is writing the script for videos, but the narrative always follows the science, not the other way around. This means that the actual script might only start to come together after a lot of time spent on the topic, when the status of scientific consensus becomes clear. At the same time, every team member checks, edits, and scrutinizes the writing of the other team members. How much planning, researching, or editing is being done on any given day mostly depends on what day of the week it is and how far along the current video is. New videos are usually uploaded on Thursday morning, and that deadline automatically leads to a rough schedule of when the script should be done, when graphics should be done, and so forth.

Talking to other people is a surprisingly large part of my work: talking to the other team members and to experts in whatever field is important for the next video. A large part of the rest of the work feels like the reviewing of research papers that I have always done as part of my research work—both as a researcher and as a science communicator, it is important to know how to read a research paper and to be able to identify the most important methodological weaknesses. Every PhD student collects a tremendous amount of experience with this: even if you're not reviewing for journals very often, you will still scrutinize the manuscripts written by your friends and colleagues and those that you review when planning or reporting your own studies. You might not realize the development of your own skills, but over the years, you'll inadvertently develop a keen eye for the overall quality of a published paper and the limitations inherent to the most important methods in your field. This skill can be incredibly important for science communication and something that you might be better trained for than someone who has studied science journalism (although, of course, that person will have other skills that go beyond those of a researcher).

What do you like most about your work?

So many things! Obviously, being part of a great team and having a good connection to your colleagues helps in any job. But what has always fascinated me about science communication in particular is the potential impact: I myself am always very happy when I learn some fascinating new fact about the world, and part of my work is now to spread this fascination to others. And my current position is a perfect opportunity for this.

One example to illustrate what I mean: statistics has been a central part of my scientific education from the very beginning. Subsequently, I noticed that this knowledge helped me even outside of science in everyday life, for example, when trying to interpret or contextualize numbers presented in the news. Over time I tried to identify those aspects of statistics that were most useful and to find ways to explain them in an accessible manner. My motivation went back to something that Carl Sagan once wrote, about how news services and politicians would have to be more honest and transparent about their numbers and statements if the overall level of understanding of statistics in the population was higher than it is. In the end, I picked the statistical concepts of median and variance as those that I found most helpful. My efforts to create accessible and understandable explanations of these concepts have led to a medium-length blog post. But even after submitting the blog post to a science blogging competition and winning third place, the readership was quite limited. Which I understood: after all, there are more exciting topics to click on than statistics. However, my current position gave me the opportunity to present the same concepts in one of the channel's YouTube videos. It's still obviously the case that statistics is a less attractive topic than others, as the resulting video got a below-average number of views for the channel. However, it still got more than half a million views, which is much more than I ever could have hoped for. And from the YouTube comments, it is apparent that a good number of people found the explanations helpful or at least not exceedingly boring. So while the size of the audience that will see my work can be intimidating and brings with it a higher intensity of scrutiny than what I am used to, it has the tremendous advantage of being able to reach out to many people and to have an impact. And if there's only one person among them that is enjoying my science communication work as much as I enjoy that of others, it will have been worth it.

And what do you like least about your work?

So far I haven't identified any real downsides, although I still have to get used to working 100% from home. It has some very real advantages, especially with a family at home as well as, of course, during a global pandemic. But I've also always enjoyed working in an office with my colleagues around me. And I think I will miss scientific conferences, although, again, the pandemic has pretty much made those impossible for the time being, so it's not really a result of me changing fields.

One element that could be considered a downside, and that I hinted at above, is the scrutiny that comes from a larger audience. The Internet can be an unforgiving place, and if you're a science communicator, there are several things that can lead to intense pressure from a large group of people: saying something inaccurate, of course, but also just discussing a controversial subject or inadvertently using terms or phrases that can be misleading or that are prone to misunderstanding. Some of these issues can be avoided by thorough vetting of every statement before it's being put on the Internet. This is something that my current team is taking very seriously so that the contents that are going out are factually accurate and phrased in an unambiguous manner. It also helps to be part of a team that includes non-scientists to avoid the tunnel vision that can sometimes plague researchers. Something else that helped me with this is the years I have spent listening to people on Twitter, which have given me at least some experience in recognizing science issues that have to be approached more carefully than others. It is tempting for a researcher to view one's own field as something that is neutral, objective, and somewhat removed from politics and society overall. But this can often be an illusion, and people might in fact have very strong opinions on one's findings, research methods, or funding structure. Twitter can be a useful training ground in this respect because users often speak their mind and can often provide constructive criticism.

All of these strategies and experiences can help avoid any potential flood of Internet-based criticism. But, and this is crucial, some negative reactions are impossible to avoid entirely, at least if you want to tackle scientific topics that are not fully settled yet or that have political implications. In those cases, there will always be some who will attack a sci comm video, or the people behind it, regardless of its accuracy. But that's just a part of working with a large-scale platform.

How do you think having a PhD has helped you succeed in your current position?

My PhD work has definitely taught me things that help me every day with my work in science communication. A big part of it is knowing my way around different research methods and data analysis. For example, I might not be working with fMRI anymore, but learning fMRI methods has taught me some common pitfalls of analyzing data that can be less than ideal in regard to its stability and reliability. And that is something that comes up quite a bit when looking through results reported from different fields.

A surprisingly important experience from that time was acting as a reviewer for publications. By that, I mean both being a peer reviewer for submissions by others and discussions within the lab of manuscripts that were still being put together. As a learning experience, it's great to review both excellent papers by seasoned researchers and texts that might be only half-done, or that were thrown together quickly for a conference deadline, by one of the more inexperienced lab members. Reviewing a range of these manuscripts, and trying to provide feedback that is as constructive as possible, is a powerful way to train one's own eye so that it can recognize gaps that others might miss. This could mean a logical leap in an argument, a misapplied analytical method, or just a phrase that is not entirely clear. I've lost track of how many reviews I have actually done or assisted with, but I would call myself an experienced reviewer, and I think this might be the strongest asset that I have when compared to science communicators with less of a research background.

A lot of people like academia because they feel it gives them an opportunity to work on a topic that they deeply care about. Do you think this is also true in your current position?

Absolutely. One reason for this is that my team is giving me the liberty to choose topics that I find important. It is no coincidence that some of the first sci comm videos that I have been involved with were on the topics of statistical testing and phrenology, two topics that are close to my area of expertise and that I believe are shrouded in a lot of mystery and misunderstanding. On the other hand, we try to stay on top regarding topics that currently interest or challenge society (there have been several on the pandemic or vaccines, for example). That means that I'm sometimes tasked with topics that aren't necessarily of personal interest to me but that are fascinating enough to cause a wider discussion. That makes it easy to care about these topics even in the absence of prior individual interest.

What might help me with this is that I've never felt tied to one particular topic and was more focused on staying within one method. In other words, I have been co-authoring papers on topics as diverse as pain, depression, aggression, fear, even all the way to topics such as dreaming and necrophilia. For some of these, I didn't have a preexisting deeper interest but rather let myself be introduced to them by seeing if I could apply my favorite research methods to study them. And this is still pretty much what I am doing now: using the training that I have to work on several different topics that I think are important.

Early in the interview you mentioned that you found that different labs have different expectations and approaches. Can you tell us more about what you were thinking about there?

In a lot of ways, labs are like families, with different expectations and different levels of interdependence. I have been very lucky in that I've always worked in labs that functioned like "good" families rather than "difficult" ones. Still, depending on your personal way of working, you could find yourself in a situation that can make it easier or harder for you to do your job. One example would be lab (or office) layout and how much privacy this affords you. I've seen it happen that people chose to work all in one large office space, even though individual offices would have been available. This phrase might be confusing to some, who would always go with the personal office and only share spaces if nothing else is available. But depending on the structure of the teams as well as their tasks, working closely together in an open space might be what works best.

A similar topic is working late nights or on weekends. Both are frequent in academia and, to a certain degree, are expressions of a work culture that allows many to structure their work day more flexibly than it would be possible in other positions. For example, if you have kids that you drop off at school in the morning, you might want to start later than others or to push some of your non-urgent tasks to the weekend or nights. If that happens out of your own volition, academia is offering you something that you might not find elsewhere. However, it might conversely be the case that nights and weekends are the times you can spend quality time with family or friends. In that situation, there might be pressure from others in the lab to work at times that will cause you stress or other problems.

There doesn't seem to be a good solution to this type of conflict. But given that lab culture differs from place to place or from PI to PI, it is at least possible to find the workspace that is most conducive to your own style, which will ultimately make you a happier and more productive scientist. This is one reason that I recommend talking to graduate students and undergrads when visiting a lab that you might be considering as a future employer: it will likely give you an impression of the overall working environment there, which should inform your decision on whether or not to work or study there.

It was great to hear about your work and advice! Thank you for telling us about your experiences, Jens!

“You Should Create a Job That You’ll Never Need a Vacation from and Turn It into a Career”

Alice S. N. Kim

Alice S. N. Kim

Abstract In our interview with Alice Kim, we discuss how an academic topic such as memory and learning can become a non-profit company focused on the application of that topic to applied contexts. After a post-doctoral position in a teaching and learning centre and not finding another organization that sufficiently aligned with her interests, Alice founded a non-profit corporation that allows her to continue doing research on topics that interest her. She does, however, also suggest that forming and growing a corporation requires ample consideration, including a financial plan, team focus, and support network. A drive to turn research into implementation can create an opportunity for yourself.

A. S. N. Kim (✉)
Teaching and Learning Research in Action, Toronto, ON, Canada
e-mail: alice.kim@tlraction.com

C. R. Madan (ed.), *Academia and the World Beyond*,
https://doi.org/10.1007/978-3-030-82606-2_6

Contents

Chris: Can you introduce yourself and tell me a bit about your current position?

Alice: About a year ago, I founded a non-profit corporation called Teaching and Learning Research In Action or TLR. I'm currently the Managing Director of TLR, which investigates and publicizes effective teaching and learning practices. Although researching effective pedagogical practices is a large part of what we do at TLR, we focus on making the results of pedagogical research more accessible to the public through non-traditional means of dissemination.

What was the focus of your PhD?

My PhD dissertation, which was a continuation of my MA dissertation, was focused on the neural correlates of how we learn at a very basic level. More specifically, I investigated patterns of electrical brain activity (or electroencephalography, EEG) that occur when we form new associations. Throughout my graduate studies, I became increasingly interested in not only how we form new associations but also how we can form new associations more effectively – in other words, how can we enhance our memory and learning. I completed both my MA and PhD in the Psychology Department at the University of Toronto, where I started my graduate studies in 2007 and completed them in 2013.

As you were finishing your PhD, what were you thinking about your career plans?

As I was finishing my PhD, I was becoming more interested in applying what we know about memory and learning, specifically how we can learn better, to real-world contexts. The first of my two postdoctoral fellowships allowed me to start exploring how memory research can be applied in educational, industry, and clinical settings. From there, I became more drawn to applying memory research in an educational context, which led to my second postdoctoral fellowship at the teaching and learning centre at the same institution where I completed my first postdoctoral position. In hindsight, my experiences at the centre as a scholarship of teaching and learning researcher had a huge impact on the trajectory of my career plans and where I am and what I'm doing today.

Your second postdoctoral position, at the teaching and learning centre, sounds a bit different than a typical postdoctoral role. What kind of work did you do in this position?

Yes, postdoctoral positions at teaching and learning centres were quite new when I held mine, and I believe they are still quite rare though growing in number. My main role as a postdoctoral fellow in the teaching and learning centre was to engage in scholarship of teaching and learning (SoTL) or, in other words, conduct pedagogical research. However, I worked closely with a team of educational developers and was given the opportunity, and was also encouraged, to explore the landscape of educational development. This led to my providing statistical support for faculty conducting research on their own teaching, as well as my facilitating various instructional skills workshops for faculty. I also continued teaching in the Psychology Department throughout my postdoctoral position at the centre. Altogether, my experiences at the centre, particularly working closely with a team of educational developers, made me more aware of several pedagogical issues. Most importantly, I've come to appreciate that there are many layers to effective teaching; when I started my postdoctoral position at the centre, my research interests were focused on implementing cognitive learning principles in course design, but now I recognize that although this is important, there are also other components that need to be considered when it comes to effective teaching practices.

Can you tell us a bit about what day-to-day life is like in your current position?

Although the impact of COVID-19 had modified my daily routine, my professional goals still coincide with leading TLR and carrying out its mission, which is to investigate and publicize effective teaching and learning practices. This involves working with an interdisciplinary team, including media producers, educators with various backgrounds, and student partners, as well as external partners (e.g. other organizations and institutions). On a weekly basis, I work with and mentor student partners and conduct and oversee SoTL and its dissemination (traditional and non-traditional means); however, a significant portion of my week is also dedicated to managing the corporation, including carving out its next steps. For example, throughout the pandemic, I've been highly involved in conducting research on teaching during the pandemic with my student partners and colleagues from other institutions, producing a SoTL podcast in partnership with a teaching and learning centre at an Ontario university, as well as consulting for other organizations and building relationships with potential external partners.

A lot of people like academia because they feel it gives them an opportunity to work on a topic that they deeply care about. Do you think this is also true in your current position?

Yes, absolutely. In my current position I'm working on topics that I care deeply about and in a way that is most fulfilling for me. At TLR, we choose the topics that we work on as a team, though all within the realm of SoTL. I can see myself doing this for the rest of my life and never getting tired of it. I like having one foot in SoTL research and teaching and the other focused on broad, widespread dissemination of SoTL research.

Can you tell us a bit more about some of the specific projects you've worked on, as examples?

Right now we're wrapping up two studies on teaching and learning during the COVID-19 pandemic. One of the studies is focused on students' experience of the shift from face-to-face delivery of their courses to emergency remote delivery, as well as their experiences of distance learning throughout the rest of the course. We had students complete guided reflections biweekly from the start of the transition to remote learning and asked them to reflect on what was helping them stay engaged with their coursework and what actions from anyone in the course they found most helpful. On the flip side, we also asked students to reflect on and identify factors that distanced them from their coursework. The study was meant to be formative in the sense that we wanted to share our findings with instructors delivering courses in the fall and winter semesters of the 2020/2021 school year, anticipating that at least some proportion of courses would continue to be delivered remotely. In addition to presenting our findings at the Canadian Psychological Association's Virtual Series (online conference) this past August, we've also shared our findings publicly on our website in written and video format. Our second COVID-19 study is an autoethnography written collaboratively by instructors and students on their experiences of teaching and learning throughout the pandemic, which we hope will provide insight on this topic from multiple perspectives.

Another series of projects that we are working on is focused on inequity in higher education, specifically on how we can break down barriers that first-generation students face in this context. The first-generation status is generally defined as neither parent nor guardian having obtained a bachelor's degree. This group of students is known to frequently encounter more obstacles that interfere with their academic success compared to non-first-generation students, making initiatives that uplift and support first-generation students very important for their success in higher education. In addition to conducting research on this topic, we are also working on a teacher-ready research review with the aim of providing educators with evidence-based recommendations for how they can design and deliver their courses in a manner that is more equitable to first-generation students. As part of our community outreach, we're also mentoring an interdisciplinary team of capstone students at York University to model a solution that enhances the success and achievement of first-generation students in higher education. We hope that all of these initiatives will help contribute to the larger objective of making higher education more inclusive for everyone.

If someone currently finishing their PhD was considering a similar position as you have now, how might they decide if it would be a good fit?

I think there are a lot of important factors to consider, and the weighting of each of them likely differs depending on one's priorities and personal circumstances. For me, being passionate about the cause of the corporation is extremely important because it's what keeps me energized and motivates me to continue working through tough times. It's also very important for me to ensure that I can lead the start-up in a sustainable manner, both in terms of the functioning of the corporation and my personal livelihood. That being said, it has been very important for me to be flexible and adaptive and to keep the bigger picture and the end goal in mind so that I can negotiate various contracts and situations to move my corporation forward. I'm learning more each day about making good business decisions – it can be very difficult, as what is best of the corporation does not always align with how I would prefer to proceed as an individual. In my experiences, it's very important to always remember that the corporation is not about me; it's much bigger than me – it's all about the cause and mission. This comes back to why being passionate about the cause of my start-up is so important for me to be able to hold my position.

If someone was interested in pursuing a similar career path, what would you suggest they do to better prepare themselves?

I would suggest that they carefully review all the relevant information about starting a corporation and that they create a solid financial and operating plan both for the corporation and themselves. Again, it is very important to make sure that the corporation is sustainable and that they have the means to lead the start-up through to maturation. I would also suggest that they make sure that they have a strong support network – both professional and personal. It is extremely important to surround yourself with people who you can trust and rely on. Choose your founding board of directors very carefully, and make sure that the members of your working team are not only skilled and talented but that they are also able to work well with the other members of your team. Lastly, you should move away from the corporation being about one person and develop systems so that it continues even when individuals transition in and out of the team.

How did you determine that starting a corporation was the next step towards achieving your goals? For instance, as opposed to working for a company that had sufficiently aligned interests.

I wasn't able to find another organization that was sufficiently aligned with my interests, particularly the focus on disseminating the results of research on effective teaching and learning practices through non-traditional means, that is, beyond peer-reviewed journal articles, conference and meeting presentations, book chapters, etc. I know of individual and teams of academics who have put forth a lot of effort to share the results of research relevant to teaching and learning with educators and have made a big impact, but I think their aims are different in that they focus on applying one or more concepts/phenomena from their own discipline to support educators with the goal of community outreach. The mission of TLR centres more on disseminating the results of recent research on teaching and learning that spans across various disciplines and through various types of mediums, including podcasts, videos, and other media. In terms of organizations, there are many that produce and support research on teaching and learning, but I'm not aware of any that also focus on the public dissemination piece.

What do you like most about your work?

It's hard to decide because I think this is always changing for me. Right now, I would say that what I like most about my work is what we do as a team and the broad scope of our work, which ranges from conducting empirical research to developing and supporting social movement campaigns and sharing information via mainstream media – all within the context of SoTL. Equally, though, it is also the people I work with – the diversity of everyone's backgrounds and expertise and how we are all brought together by our beliefs and concerns for various topics within the broader context of education to work collaboratively. Needless to say, everyone I work with is very special. I also feel very gratified when the results of our research are applied through others' teaching and/or learning practices – this is extremely meaningful to me because I believe that education can be life changing and at the heart of solutions to important societal issues.

What do you like least about your work?

What I like least about my work – that the scope of our work is limited by lack of time and resources. I think it can be challenging to decide what projects will be pursued by our team given that we have broad interests and we're still in the very early stages as a corporation. I look forward to growing as a corporation so that we will have more resources to take on more, or larger-scale, projects. Currently, however, we are limited by our available resources, but perhaps this is the sentiment of most others as well in different positions and how I will always feel no matter how large TLR grows.

Are there some aspects of academic research that you miss?

I think I'm still closely linked to academic research via the SoTL research I do through TLR. Although SoTL is academic, it is not highly celebrated as a prestigious field of academic research (at least that is my sense and many others in the field), but it should be. Part of the reason I started TLR was to bring more attention to SoTL and the impact it can have on teaching practices and more generally education. I think my current portfolio essentially includes almost everything I did in my postdoctoral position at the teaching and learning centre and more. To recap, I think academic research is still a part of what I do, it is just that the context is now a bit different.

Based on your journey, what is some advice or suggestions you would want to pass on to someone who's currently finishing their PhD?

The best piece of advice I was given is that you should create a job that you'll never need a vacation from and turn it into a career. On top of this, I would say that although there are definitely systems in place, there is still space and ways to work outside of traditional spaces and positions to carve out a career that works for you. If you find that there is a dream job out there waiting for you, then you should go for it! However, if everything is less than perfect (or what you would be happy with), don't be afraid to create something new. In other words, if you don't fit the mould, try creating something that fits you.

Is there anything else you'd like to tell someone reading this interview?

The only thing I would mention at this point is that, needless to say, everyone's experiences are different. I think it is helpful and informative to read about the experiences of others, but it is also important for the reader to remember that their life circumstances and the factors that they need to prioritize may not match those of the individuals sharing their experiences in this book. Generally, I would say that the reader should move their lives in a direction that works for them and their specific circumstances.

Thank you so much for telling us about your work and career path, Alice!

"A PhD Has Helped Me Learn How to Think Through Problems"

Aaron Moss

Aaron Moss

Abstract In our interview with Aaron Moss, we discuss how his post-PhD career focus changed from research to teaching to industry. Aaron's experience using online research platforms for his graduate research made him a great fit when he applied for a position at CloudResearch, a company that connects behavioral researchers to online research participants. The job involves keeping up with current research and publishing research while also helping other researchers conduct their studies. Aaron reminds that while graduate school is associated with many uncertainties, it is important to figure out what your goals are and work toward those during graduate school.

Contents

A. Moss (✉)
CloudResearch, Queens, NY, USA
e-mail: aaron.moss@cloudresearch.com

C. R. Madan (ed.), *Academia and the World Beyond*,
https://doi.org/10.1007/978-3-030-82606-2_7

Chris: Can you introduce yourself and tell me a bit about your current position?

Aaron: I'm a Senior Research Scientist at CloudResearch, a company that connects behavioral scientists with online research participants. Most of my days are spent thinking about ways to improve online research methods, sampling, and data quality.

Together with the members of our research team, I work to publish papers that help researchers understand the online research landscape. With every project we conduct, we're interested in demonstrating how researchers can do things better or more efficiently.

In addition to research, I write a variety of content for our website and work to build relationships with the researchers who use our services. I enjoy being in a position where I can see how research is conducted both in academia and industry.

CloudResearch is based in Queens, NY, but I live in Albany. Originally, I am from Indiana. When I'm not working, my wife and I enjoy cooking, reading, following sports, and taking our young son on hikes or to the park.

What was the focus of your PhD?

I received my PhD in social psychology in 2018. I worked with Laurie O'Brien at Tulane University in New Orleans, Louisiana. My research primarily focused on prejudice and intergroup relations. More specifically, I was interested in how people make judgments about which behaviors or policies constitute discrimination and how people view their own prejudice.

Some of the projects I worked on investigated how moral concepts like intent and harm shape people's judgments of discrimination. These projects were fun and used some uncommon methods like coding of archival data. Other projects I worked on looked at what factors lead groups to express prejudice. But most of the projects I led were interested in how people think about their own prejudice.

Early in graduate school I ran some studies investigating whether people engage in motivated reasoning and other defensive behaviors after being presented with evidence of their prejudice. When these studies didn't work out as I'd hoped, I started to investigate what we think of people who admit or deny that they have engaged in prejudiced behavior. Then, finally, my dissertation investigated how social norms might be leveraged to encourage people to admit prejudice.

As you were finishing your PhD, what were you thinking about your career plans?

I worked on my dissertation for about a year and a half. During that time, my career aspirations changed quite a bit. Initially, I was like most graduate students in that I was hoping for an academic job. At some point in my program, I realized that any academic job I may land wasn't going to be research focused; I just didn't publish enough or have the high-impact publications needed to be competitive. So I became interested in a teaching-focused job.

However, my situation was complicated by life circumstances. During graduate school, I got married to another academic. My wife was a few years ahead of me in school, so by the time I was looking for a job, she had completed a post-doc and started a tenure track job at a liberal arts college. During this time, we lived apart for three years.

I applied to a handful of academic positions in the geographic area where my wife was employed. After I didn't hear anything from those jobs, I began thinking about industry. I quickly found that many people with PhDs were working in market research, UX research, and government jobs. UX research seemed especially appealing because these jobs were often with tech companies and presented some interesting opportunities for research. Most tech companies, however, were also located in places I couldn't move because of my family. So I focused my job search on market research positions while keeping the door open to other opportunities. After just a few months of searching, I found a listing for my current position and applied.

Was there anything that you did to better prepare yourself for a UX research job?

I started taking steps to explore UX careers but didn't get very far. My non-academic job search lasted about two months before I started my current position. During those 2 months, I conducted some informational interviews with people in UX positions and spent time reading about the methods commonly used in UX studies. There are a lot of social psychologists and other behavioral scientists employed as UX researchers, so it isn't hard to find people to talk with.

Can you tell us a bit about what day-to-day life is like in your current position?

Day-to-day life is busy. CloudResearch is a small company; we have less than 30 employees. Our size means we're often working on many different projects and juggling many competing priorities.

Most of the time, these competing priorities mean my days are a lot of fun: I get to focus on a handful of different problems, switch gears a few times throughout the day, and interact with different teams within our company. Other days, these competing priorities can be stressful. Fortunately, the good days greatly outnumber the stressful days, and watching our company grow to serve the needs of more and more researchers is its own sort of reward.

A good day for me begins with some organization. I typically handle emails, organize my to-do list, and participate in our daily stand-up call. For the next few hours, I may work on a research paper or write some website content. Next, I may spend an hour or two in internal meetings or calls with clients, and then I return to my writing projects, research tasks, or more emails. I try to balance working with others with having some time to myself to think and write.

How do you think having a PhD has helped you succeed in your current position?

One way having a PhD has helped me is by helping me learn how to think through problems. I couldn't point to when this happened, but somewhere along the way to my PhD, I got a lot better at analyzing problems and thinking my way to a solution. The ability to confront a new problem, analyze it, and find a solution is important to what I do, and my time in graduate school helped me develop this ability.

Another way having a PhD has helped me in my current position is by giving me time to further develop important skills and knowledge. When you're a graduate student, you have a lot of unstructured, free time (I completely understand if it doesn't feel that way to current graduate students). During this free time, no one is telling you what to do. You can learn R, study advanced stats, become a skilled writer, or develop any number of other skills that are an important part of being an academic. During my free time, I focused on writing.

I was interested in writing before graduate school, but during graduate school, I had more opportunities to practice and more resources. I read several books about writing, and I participated in a writing workshop my department put together. I also volunteered to contribute blogs to an online psychology magazine in addition to all the writing I did for classes and the articles I was working on. Collectively, these activities helped me develop as a writer, and writing is probably the most important part of my job at CloudResearch.

The other piece of knowledge I developed, unintentionally, was an understanding of online research tools and participant platforms. I started graduate school in 2013. Most academic researchers began using platforms like Mechanical Turk in 2010 or 2011. This timing meant that as researchers in my area were adopting new tools and learning how to run a wide variety of online studies, I was learning about these things too. I didn't realize until much later how information that was common knowledge among people in my field was actually pretty esoteric. Once I began interacting with researchers from industry and other fields, I began to appreciate how quick social psychologists were to adopt some of the online tools that emerged over the last decade and how I benefited from that collective knowledge.

If someone currently finishing their PhD was considering a similar position as you have now, how might they decide if it would be a good fit?

An internship or some other hands-on experience is probably the best way of finding out if a job seems like a good fit. Short of that, however, I think there are two ways to assess whether you would be happy in a particular job.

First, talk to people. Ask people who are currently in positions you'd be interested in what their days are like, what they like about their job, and what they find challenging. Talking to other people is one of the best ways to learn about things outside of your experience.

Second, I think you have to be honest with yourself and evaluate how you want to spend your days. What kinds of activities are going to make you happy day after day? What kinds of activities are going to wear you down or demand too much from you?

I was interested in an academic job for most of graduate school because I liked doing research. When it became clear that any academic job I may get would be focused on teaching more than research, I signed up for a teaching pedagogy course and taught my own class. Although I enjoyed some aspects of teaching, there were many things I did not like. I continued to pursue teaching positions for a while because I wasn't aware of the alternatives. If I was honest with myself then, I probably would have started exploring non-academic positions a year or so before I did.

So I think it's important to assess what kinds of activities you want your days to be filled with and then ask whether a given position will deliver that.

What were some factors that led you to not pursue a teaching career further?

Teaching is a tough job. I find it draining to be "on" in front of a class for long stretches of time. I taught one course in graduate school, and I struggle to envision what it would be like to teach between two and four courses per semester and be in front of a class for most of the week.

In addition to what teaching demands from me personally, I also realized that a teaching-focused position wouldn't allow me to spend my time doing the things I find most fulfilling. I applied to graduate school and pursued a PhD because I like research, not because I like teaching. While I found it enjoyable to work with students and give people the resources to pursue things they are passionate about, I also felt that teaching wouldn't sustain me over the long haul. I eventually realized that I'd be much happier if research was the primary focus of my job rather than something I was expected to do on the side.

If someone was interested in pursuing a similar career path, what would you suggest they do to better prepare themselves?

I'd recommend making sure you have a strong foundation in research methods and that you're able to communicate well. I didn't do anything special to prepare for my current position, but I draw on nearly all the things I invested time in learning while in graduate school, and near the top of that list is probably communication skills.

If you work in industry, you will have to explain your research and its importance to people who do not share your background. Other members on your team or stakeholders within your company will sometimes look at you to explain complex things in a way that they can understand. Being able to translate your research so that it's accessible to other people is really valuable regardless of whether you work in academia or industry, so it's a skill worth developing.

What do you like most about your work?

I like that I get to write papers for publication in academic journals. When I started looking for a job in industry, I assumed I wouldn't be doing any more of that type of writing. But at CloudResearch, we see a lot of value in communicating to scientists in peer-reviewed outlets. It isn't always easy, but I really enjoy telling a story with data and putting together a polished paper.

I also like that I get to learn lots of new things. Online research methods evolve quickly, and as technology continues to improve, researchers are going to look for new ways to collect higher-quality data about human behavior. I like being at the forefront of this work and thinking about ways to improve research.

I also like learning about the breadth of research people conduct. In just the few years I've been at CloudResearch, I have learned a lot about how researchers in various academic disciplines conduct their work and the types of things they care about. I also get to learn about how researchers in industry go about their work. If I stayed in academia, I think it's likely I would not have gotten such a broad perspective. In academia, it's easy to stay siloed within your department or your area, so I value the breadth of research I get to learn about in my current position.

And what do you like least about your work?

It sounds cheesy, but I don't like letting people down. We facilitate a lot of research projects. Sometimes, we just give people the tools to conduct studies themselves, and sometimes, we play a more active role in helping collect the data. Our clients' research is important to them because they have a lot of time, money, and personal commitment wrapped up in their work. The research is important to us because we exist to make that work easier and more efficient. Despite everyone's best efforts, there are times projects don't work out.

Sometimes, there are issues with data quality or our ability to recruit certain samples. At other times, researchers may want to use a task that isn't feasible, or the data needs to be gathered under a tight timeline, which creates strain. Anytime we can't solve a problem or ensure the client is completely happy with their project, it is unpleasant and that is the part of my job I like least.

A lot of people like academia because they feel it gives them an opportunity to work on a topic that they deeply care about. Do you think this is also true in your current position?

I've always had broad interests. I admire academics who find one topic that they're really interested in and carve it out as their own niche. I've never had a topic like that. Even though I cared a lot about the topics that led me to graduate school, I was interested in a wide set of questions about human behavior.

In my current position, I get to work on things that I find enjoyable and that are general enough to satisfy my curiosity. In some ways, I get to do research about research, and I find that really interesting. Sometimes, we're able to look at data about how thousands of researchers across various disciplines run their studies. Then, we can use that data to make recommendations about how to do things differently. When we're really lucky, we get to see people implement those recommendations into studies that investigate all sorts of interesting questions about human behavior. That is pretty exciting.

Also, because my career is academic adjacent, I have the opportunity to collaborate with people to pursue questions I am interested in but that aren't related to my work at CloudResearch.

Another reason many like academia is that they feel it provides them with more freedom than they think they would get in other positions. How much freedom do you feel you have to work on what you think is interesting?

I have quite a bit of freedom, but within boundaries. As an academic, you can run a study on nearly anything you want as long as you have the resources or funding to do it. The projects I work on have to have some relation to our business, but even within that space, there are lots of possibilities.

Our research team makes decisions about which projects to pursue in a pretty collective fashion. There are more projects to pursue than we have people to carry out the work. It probably helps that I'm interested in a lot of things, but I can't think of a single research project we've pursued that I didn't want to work on.

Every job has some aspects that may not be exactly what you want to do. I don't know how many academics *want to* attend faculty meetings. My job has some things I do because they are part of my job and they need to get done. But even with those things, I have a lot of freedom to work on them the way I want to.

Based on your journey, what advice or suggestions do you want to pass on to someone who's currently finishing their PhD?

If you're finishing a PhD program and you're thinking about a job in industry, take some steps to prepare yourself. Even though you may have little exposure to what people with PhDs do in industry, realize that most people with PhDs are not employed in the academy, so it isn't like you have to reinvent the wheel. There are books and websites solely devoted to helping people with PhDs transfer to industry. Find these. You or the people you know may have contacts who work in industry. Talk to them. If you're finishing a PhD program, then you're clearly smart and you know how to persevere. Getting the job you want in industry is a matter of charting a course and taking the steps to get there.

Is there anything else you'd like to tell someone reading this interview?

Since getting my PhD, I've thought a lot about how uncertain life is for graduate students. As a graduate student, you're not making much money, you don't know what job you may get or where you might live after school, you're probably somewhat unsure of your ability and status within your profession, and together, all that uncertainty creates a lot of stress—not to mention stress from other serious things like mental health, climate within your department, family problems, or other issues.

For most people, I think the uncertainty fades after finishing a program. As you settle into a job, you're more financially secure, you have closure on where you're living and working, and you're hopefully more comfortable as a professional. In other words, a lot of situational stressors can change pretty quickly. And, then, they are replaced by something new.

When people would tell me congratulations after I got my PhD, I would joke and say, "Yeah, now all I have to do is figure out how to live 50 or so years of life." It was a joke, but it was also true. After grad school ends, you have to figure out what's next and how to live your life. Hopefully, the work you did in graduate school allows you to pursue work that you find meaningful and that makes you happy. Remembering what your goals are and what you're working toward while in graduate school is important because remembering is what helps you get through.

Thank you so much for telling us about your career and journey, Aaron!

"A PhD in Industry Is a Revered Qualification"

Joseph M. Moran

Joseph M. Moran

Abstract In our interview with Joe Moran, we discuss how he transitioned from academia to data science, including some of his missteps along the way. Joe provides some insights from his experiences on how PhD training can be applied in an industry context, such as in user experience (UX) research. Moreover, Joe provides direct and helpful recommendations on how someone can become better prepared to engage in data science or UX research. PhD students should do their best to discuss potential career paths with those in different jobs, both in and out of academia.

Contents

J. M. Moran (✉)
Rapid7, Boston, USA, MA

Google, Cambridge, MA, USA

C. R. Madan (ed.), *Academia and the World Beyond*,
https://doi.org/10.1007/978-3-030-82606-2_8

Chris: Can you introduce yourself and tell me a bit about your current position?

Joe: In my current work, I lead a team responsible for customer research at an Internet security company. Whenever people have a question about our customers they don't know the answer to, they can reach out to us, and we can help them get that answer.

Most of my work involves interfacing with other teams to ensure that the work we are doing remains focused on business goals and the goals of the product and marketing teams with whom we work. I then direct my teammates toward projects and efforts that will enable us to deliver against company goals and against the needs of the organizations we work with. Briefly, we conduct two kinds of research: (1) research that supports the teams that build our products, which is focused mostly on qualitative studies about user needs and product issues (UX research), and (2) research about how our marketing efforts are working with individuals who are not yet customers. This work is more typically quantitative in nature, looking at things like web analytics and conversion rates, for example.

What was the focus of your PhD?

I completed my PhD in the Department of Psychological and Brain Sciences at Dartmouth College in 2006. My PhD work was in the field of social neuroscience, focused on the neural representation of the self and others. This work began by investigating how the brain encodes information about the self – when I ask you to reflect on whether individual items are related to you, which parts of the brain predict whether you will later remember having seen that information when I ask you to recognize those items?

I then continued in this vein, further looking at the boundaries of the roles of the default mode structures in self-reflection and self-relevance. It was exciting work in a cross-disciplinary way, picking up skills and knowledge from cognitive neuroscience, statistics, programming, and social cognition along the way. One of the things I particularly liked was getting to attend conferences from very different scientific traditions while learning the particular languages of the subfields I was wandering into.

As you were finishing your PhD, what were you thinking about your career plans?

My initial considerations, both during and toward the end of my PhD, were that I would continue in the same path as my mentors, onto a tenure track position at a research university. In fact, at Dartmouth at that time, that was very much the expectation of the program on its students. We were told, both indirectly and directly, that we should view success as a faculty position at an R1 university and any other career path as less-than. This certainly influenced my feelings about looking outside academia at the time. My initial steps were to take a postdoctoral position at MIT, where I continued and expanded on my PhD work by looking at how self-reflection, and its neural instantiation, might differ in disorders like autism and schizophrenia and how it might change as we age.

How have your career plans changed as you've continued on to your current position?

During the financial crisis of 2008 and beyond is when I was seeking academic positions. There was significant competition for very few assistant professorships at the time, and so I ended up taking a second postdoc and continuing my search. As I ventured further into the job market, year on year, several nagging thoughts began to emerge that maybe this wasn't the path I wanted to be on. The yearly grind that I would have to potentially move my wife and young children to a random city far across the country became harder and harder to face. And I was beginning to wonder for what I would be doing that? As my friends became assistant professors around me, and I started to learn about the tight focus and continuous grant writing that a tenure clock demanded, I began to really examine whether those things were what I wanted, especially given the aforementioned social costs associated with pursuing such a career.

As I thought more deeply about what I wanted to spend my time doing, it became clear that a path to industry might be the right one. I enjoy solving problems, I enjoy evolving challenges, and I enjoy working on things that have rapid impact on the environment around me. Academia did not seem a place where I could engage in two of those three things.

I started to learn more about industry roles like "data scientist" and "user experience (UX) researcher" from friends who had made the switch from academia, from introductions and from coffee chats and information interviews alike. The process to becoming competitive for positions like this, coming from academia, involves a lot of shedding assumptions about how the work is done and what kinds of activities and postures get rewarded in industry. It took me about 12–18 months before I was able to get my CV into the shape that would get past the filters for industry jobs and then another 6–12 months of phone screens and initial interviews before figuring out how to speak like an industry data scientist or UX researcher. There were many missteps in those initial outreaches to companies, where you could almost hear the door closing while the recruiter was still on the phone.

If you don't mind, can you tell us a bit about some of those missteps?

Most of the difficulties I faced in the early parts of my search were related to not being familiar enough with specific techniques in data science and machine learning. At one company, I failed a phone screen because I hadn't worked on collaborative filtering for content recommendation; at another, I didn't make the cut because I hadn't come across multi-objective optimization before. I realized that this confusion was a two-way street: recruiters new to finding data scientists were over-indexing on single methods on job descriptions, and I was over-indexing on what I already knew, without joining the dots between these techniques to present as more well-rounded. As I learned to navigate these complexities a little more successfully and make it deeper into the selection process, other things tripped me up. At one company, who was looking for their first data scientist, I readied my Data Science 101 presentation, turned up at their HQ in a different city, and proceeded to get laughed out of the room by some well-seasoned machine learning engineers – I had not realized I was applying to a smaller subsidiary of a larger company who in fact had a very well-established team and was looking for a similar level counterpart to run data science in simply a different office.

All of these missteps and failings produced learning opportunities however. I began to practice different techniques on nights and weekends, working on data science problems in my spare time, until I felt confident enough that I had a broad enough understanding where I could come up with good-quality solutions to hypothetical problems posed during on-site interviews.

Can you tell us a bit about what day-to-day life is like in your current position?

In my current position, as manager of a customer research team, I have moved more toward the qualitative side of the equation. My team consists of six full-time researchers, ranging from psychology PhDs to folks with marketing backgrounds and beyond. Day-to-day life for me typically involves one-on-one meetings with my team members: opportunities to dig deeper on their current projects and needs and less frequent development conversations around their goals for personal growth and exposure to new methods and techniques. Within those conversations, I get an opportunity to mentor junior folks and support the more senior members of my team in expanding their horizons beyond the focus of my department.

Personal and team success in industry is all about relationship building. While academic science is collaborative and requires working with others, those roles are more easily constrained by the nature of the work. A postdoc manages the day-to-day work of a research assistant while perhaps mentoring a grad student on a new analysis technique. In industry, the success of a manager really depends on how well they can communicate the role of their team in helping others to succeed. So most of my day-to-day life is relationship building and maintenance with managers and directors from other departments: Experience Design, Engineering, Product Management, Marketing, Customer Success, Onboarding, etc. My role is to "sell" the results of my team's primary research to encourage improvements to products and processes that result in reduced financial risk to my company. A fair bit of my work involves evangelizing the existence of UX research as well – industry is fast-paced, and each team you interface with has varying degrees of facility with all the ins and outs; the product team of a company we recently acquired had almost no knowledge of the role of a UX researcher, and so we've had to build that relationship from the ground up.

Other day-to-day things involve cross-functional alignment. This means realizing where there are process gaps between teams (one of the benefits of being in UX research is you get to talk to everyone) and then creating ideas to improve those processes and bringing those ideas to the teams involved. Lots of slide deck preparation and presentation and then iterative work to zero in on what would make the most sense. Here, too, collaboration is critical – you need to approach the work with humility, recognizing that quality ideas can come from anywhere and that the real work is in capitalizing on those ideas together to ensure everyone's success.

How do you think having a PhD has helped you succeed in your current position?

The first and most direct way that a PhD in psychology has helped is domain specific. Learning about cognitive and social psychology – heuristics and biases, fundamental attribution error, framing effects, attitudes to risk, etc. – has enabled me to apply these principles to the work that we do every day, both in thinking about how my colleagues react to events and how our customers interact with our products. One of the things we make is a risk score – I was able to contribute to early discussions with our data science and product management teams a perspective around how people perceive risk and how framing motivates risk-seeking and risk-aversive behavior. A background in statistics has come in handy when working with marketing research folks – teaching them how to build confidence intervals, how to run simple inferential tests, what representative means, what normally distributed means, etc.

Beyond these domain-specific things, a PhD has enabled me to bring a depth of thinking and rigor to an environment that rewards breadth and speed. Industry professionals are excellent at moving quickly – spinning up ideas and testing them out, moving forward on relatively little feedback. Having a PhD enables me to do the work of knitting together those observations over time into a coherent framework and then reinserting that knowledge into the development process when appropriate, thus providing more context and depth for future ideas and experimentations. But this didn't happen immediately – initial attempts to bring this deeper perspective in my first industry position brought blank stares – you have to be able to move at industry pace while building up deeper insights. There's no long pause button where you get to go away into the cave and come back with a fully formed theoretical viewpoint.

Finally, a PhD in industry is a revered qualification. It can open doors in executive discussions and in customer-facing engagements. Very few people in industry have PhDs, and the title definitely lends a certain amount of gravitas. But it is important to bring the rigor and ethical stance with you when you earn that gravitas; just because you have PhD after your name doesn't mean that whatever you say is right – you need to keep your instincts sharp and make sure you don't stray beyond what you know is true.

If someone currently finishing their PhD was considering a similar position as you have now, how might they decide if it would be a good fit?

Three methods that can help you learn about what it is like to work in an industry UX research position are coffee chats, informational interviews, and shadowing. I arranged them in order of formality; coffee chats are a simple method to gain awareness of what it's like to work in a given industry position, from a current practitioner. These can be helpful to increase the size of your network and gain an ally who can help you figure whether job opportunities you spot might match your skills and experience, for example. Informational interviews are a little more structured and involve asking a series of predefined questions (lots of templates and ideas online) designed to help you decide if this sort of work suits your style, personality, and goals. Finally, shadowing can be great to get a sense of what the actual day-to-day life of a UX researcher in industry looks like.

As you engage in these ways, think through how you like to spend your time each day. Are you deep in the data, exploring and writing code? Are you thinking more about what we want to study and how we might go about that? Do you prefer spending your time mentoring more junior folks? Having a sense of what you like and don't like about your current work will help you to make sense of what you would prefer in an industry setting.

If someone was interested in pursuing a similar career path, what would you suggest they do to better prepare themselves?

My experience has been in data science and in UX research, so I'll limit my comments to those domains.

Data Science
Learn one programming language really well, whether Python or R. Depending on the positions, you might be a "type A" or a "type B" data scientist. Type A is for analyst, and that is someone who writes code to connect to data sources, ingest data, clean and tidy it, run descriptive stats and create visualizations, and then create inferential statistics and models to, for example, determine whether an intervention worked or forecast future states. Type B is for builder, and that is someone who is creating (typically) machine learning models that need to run in production and interface with all the other data pipelines and production systems that companies use. Generally, but not always, type B data scientists will work in Python, whereas type A data scientists might use R, Python, or a mix of the two.

Data scientist interviews typically involve your knowing why, when, and how to implement particular models or algorithms, with discussions about projects you have completed where you did so. They will also use hypothetical data situations to get a sense of your creativity in coming up with plans for modeling the kinds of data that their company might collect. To relate previous experience to their industry perspective, focus on why you made analysis choices you did, how you worked on the data, and what the end result enabled.

UX Research
Industry UX research teams want to know that you can choose successfully among the wide array of user research methods available to creatively answer given industry

challenges. They'll look for you to show how you have been creative with methods in the past to figure out how to learn what the product teams need to know and will also pose hypothetical questions about situations that might face their teams currently. Spending quite a bit of time on figuring out how industry terms relate to academic science terms, and being comfortable translating between the two, would be of great benefit. For instance, contextual inquiry, a common industry method, means nothing more than watching how people complete tasks in a natural setting and (maybe) occasionally asking probing questions about what it is they are trying to get done and how. How has your academic work prepared you to do that kind of work in industry? I'll bet psychologists can think of many ways that a keen eye for observation has helped them narrow down experimental hypotheses.

Further, building up a portfolio of work you've actually done can be beneficial too. You might conduct a heuristic evaluation of a popular consumer website or run a usability study on some workflow in an application that you like to use. For these, one can recruit friends and analyze the data independently. Hiring managers like to see your process and learning, which can easily be captured by doing a project like these.

A lot of people like academia because they feel it gives them an opportunity to work on a topic that they deeply care about. Do you think this is also true in your current position?

In short, both yes and no. It depends on how you define "topic," and it depends on how you define "deeply care." In industry, you're not free to pursue for long periods of time a particular topic through iteration and careful testing of parameters over years. That much is for sure. However, I would argue that in academia there is actually quite a bit of constraint on the topics one can choose to work on. You need to choose something that funding agencies care about, that hiring committees value, and that departments see fitting in to their broader community. You can't simply commit to a huge research program on your esoteric topic if you can't convince others to give you the space, funding, and students with which to do it. And the same is true in industry – we're definitely constrained to focus on things that the broader company thinks are important. The level of constraint in industry is just tighter compared to academia.

Secondarily, I mentioned how you define topic. Some folks are driven by a particular question or topic their whole careers. For those people, I think academia is likely the right place. But many are driven by the intellectual challenge of problem-solving. Still more are driven by connecting their work to more visible or more rapid impact. So for those kinds of people, topic is less important, and it might make sense for them to consider a career in industry.

Another reason many like academia is that they feel it provides them with more freedom than they think they would get in other positions. How much freedom do you feel you have to work on what you think is interesting?

This is an interesting topic too. As I mentioned previously, I often think there are more constraints around topic in academia than people realize. Certainly in psychology, if you had, for example, not been using neuroimaging over the last 15 years, it may have been more difficult to secure funding and attract top students. While you are still free to investigate your topic of choice, it is a more difficult proposition if that topic is not "fashionable." While I am not free to choose anything I like to work on, I can go and sell my labor elsewhere if I don't like what my company is working on. And there is a lot more mobility in industry than in academia. Because of the conveyor belt nature of academic positions, you don't see people switching labs too often, whereas in industry, in the current employee-leaning market, people are free to switch companies quite easily. This enables you to work in quite a few different domains over a relatively short space of time, and many people expressly avail themselves of these opportunities; it's not just a theoretical freedom.

Another measure of freedom is time; I've seen quotes like "In academia you are free to work whichever 60 hours in the week you choose." And yes, it's possible to organize your time slightly less around a traditional workweek in academia, but in industry, there rarely comes a week where one needs to put in any time in the evening or on the weekend.

A final measure of freedom is financial; it is well known that industry positions tend to pay better than academic ones. If you seek financial freedom and have the discipline to save aggressively, it follows that it becomes easier to secure financial independence in an industry career.

What do you like most about your work?

The opportunity to solve difficult problems that underlie a need my company has. The chance to contribute to initiatives that change the business and its approach and the time horizon to see those initiatives succeed or fail. Getting to collaborate with and learn from specialists in all aspects of what the business does. I also collaborate with team members from across the globe, so getting insight into conditions in different parts of the world is fascinating too.

What do you like least about your work?

There is very little I dislike about my work. After quite a bit of thinking, I'd say the only difficulties emerge when I have to deal with competing priorities for different teams or shifting expectations from senior management about goals and progress. Things do shift quickly in industry, which is exciting but sometimes leaves you having to scramble to re-focus efforts or spin up new things quickly. That can definitely be a challenge.

Based on your journey, what is some advice or suggestions you would want to pass on to someone who's currently finishing their PhD?

Well, first up, I think everyone's journey is different, and any advice you receive should be taken with a grain of salt based on the idiosyncratic path the advisor has walked. There's a lot of survivorship bias in academia and beyond, and so what "worked" for me or anyone else is also impacted by luck and privilege – positive and negative, seen and unseen – and one should always be mindful that the neat stories we tell ourselves about how we got here often conveniently ignore the role that both randomness and structural societal factors play.

That being said, I think that current PhD students who are considering careers both in and out of academia should spend some time getting to know people who are currently doing the jobs they are considering: postdocs and senior UX researchers, data scientists, and the like. Make a list of people you know who know someone in the particular industry, and start reaching out for the kinds of coffee chats and informational interviews I mentioned previously. This is a great way of getting a peek behind the curtain into what it is like to work in a given industry or profession and will go some way toward helping you understand whether your skills, interests, and experiences could be a good match for the things you see people doing in these jobs today.

Beyond that, if you are considering a career outside academia, start reading job descriptions now and figuring out what the terms mean. Skills you have will very likely match what hiring managers are looking for, but if you use academic terms to describe them, you'll risk having your application overlooked by either an algorithm or a busy recruiter who doesn't know that "structural equation modeling" is a technique that could explain why consumers in their industry behave the way they do.

Finally, all the skills building in the world will not matter if you don't build your network. Connect with people in LinkedIn, Twitter, and elsewhere, and get to know as many people as you can inside your chosen industries. Many jobs get filled via referrals, and the only way to make those happen is to get a connection to someone on the inside.

Thank you so much for sharing your experiences and insights!

“No Matter How Specific Your Interests, There Are Options Out There”

Alison Caldwell

Alison Caldwell

Abstract In our interview with Alison Caldwell, she tells us about her graduate research as well as her accidental hobby, Neuro Transmissions. What initially began as helping make some fun promotional videos for the neuroscience department’s graduate program helped start a YouTube channel that now has over 100 videos, teaching introductory neuroscience topics. This was all done concurrently with her PhD studies. Alie gives some advice on how to get started in science communication and talks about her current position working as a scientific public information officer for her university.

Contents

A. Caldwell, (✉)
University of Chicago Medicine, Chicago, IL, USA

Neuro Transmissions, Chicago, IL, USA

C. R. Madan (ed.), *Academia and the World Beyond*,
https://doi.org/10.1007/978-3-030-82606-2_9

Chris: Can you introduce yourself and tell me a bit about your current position?

Alie: I'm Alie Caldwell, a neuroscientist, science communicator, and writer. I recently completed the Bigelow Memorial Science Communication Fellow at UC San Diego Health, and I'm now the senior science writer at University of Chicago Medicine, where I work in branded communications. The unique Bigelow Science Communication fellowship is a collaboration between three different media relations departments on campus – Health, Oceanography, and Engineering. I spent 3 months embedded in each of those departments, writing press releases and news features, creating podcasts and videos, and working with media relations staff to develop new ways to tell stories about university research. Because of my background in informal science outreach, I also consulted with the departments on how they can leverage their digital media channels to improve their public outreach. Now, at UChicago, I produce stories about the biological and medical research happening on campus.

I also have a side hustle that started during graduate school and has continued for almost 5 years. I'm the co-creator, writer, and host of Neuro Transmissions, a YouTube channel all about the brain, where we create educational videos on all kinds of topics in neuroscience and psychology. This project has led to a number of professional opportunities, including invited talks and the opportunity to write a non-fiction pop science book. Through my YouTube work in grad school, I was able to build a strong community to support my research and creative work and connect with those many outside opportunities.

What was the focus of your PhD?

I received my PhD in August 2019 from UC San Diego. I chose UCSD because of the multitude of talented PIs I could possibly study under and because I could see that the students there were enthusiastic about their research but also seemed to have hobbies and interests outside of the lab. As luck would have it, I ended up rotating with and then joining the lab of one of the PIs who interviewed me there – Nicola Allen at the Salk Institute for Biological Studies. Nicola was trained as a postdoc by Ben Barres, the godfather of glia, at Stanford University, and her lab is focused on understanding the roles of astrocytes in the brain and particularly how they influence synapse formation, maturation, and degradation.

For my dissertation research, I focused on understanding the proteins produced by astrocytes during a key time point in development – at postnatal day 7 in the mouse cortex, when astrocytes are known to be involved in neuronal outgrowth and synapse formation. I adapted existing protocols to isolate astrocytes and neurons in vitro to produce age- and region-matched cortical cultures in order to profile the astrocyte secreted proteins and test the specific effects of individual proteins on neuronal development. In particular, I was looking to see if there were differences in the protein secretion profiles of astrocytes from genetic neurodevelopmental disorders when compared to wild type and to see if there were similarities among the different genetic disorders. Using serum-free culture conditions, mass spectrometry, and RNA sequencing, I identified over a thousand proteins that show differences in secretion in the disorders compared to wild type and further investigated two proteins for their possible effects on neuronal outgrowth. My work showed that these proteins – BMP6 and Igfbp2 – both lead to stunted outgrowth in neurons in vitro and that blocking those proteins in media containing disordered astrocyte-secreted proteins can partially rescue the outgrowth defects associated with the conditions.

I didn't publish any first-author papers during grad school, which is a bit of a sore spot, as my advisor had different publication goals than I did. As a result, this work is still unpublished, though it is currently in revisions; we had some very positive reviews and are working to complete the requested additional experiments before we resubmit. I am now consulting on the project and will remain first author, while others in my advisor's lab will complete the additional work.

Can you tell me more about Neuro Transmissions? How did the idea to make a YouTube channel get started?

I sort of fell into YouTube by accident! My graduate program had started a tradition before I arrived on campus of making music videos to promote the UCSD Neurosciences Graduate Program Social event held during the Society for Neuroscience meeting whenever it happened in San Diego. It started with Lady

GABA's "Poster Face" in 2011, and in my first year of grad school, they were creating one called "Get Data," based off of Daft Punk's Get Lucky.

My partner has long had an interest in filmmaking and music, so when I learned about the project, I asked if the student group could use any help. They were eager for anyone who had videography and film editing experience, so my partner jumped on board, and the ensuing video went mildly viral in the neuroscience community.

I thought it would be a one-off thing, but then a friend sent me a link to the Society for Neuroscience's annual video contest and asked if we were interested. My partner and I decided to enter, and I wrote up a script. We filmed the video while we were on vacation in the Rockies and ended up winning second place! But more than winning the award, we had a lot of fun collaborating on the project, and it got the wheels turning for me. I'd always considered myself a writer, but I'd only ever done creative writing and mostly fiction. This was the first time I'd ever attempted nontechnical science writing, and I found it really exciting. Plus, at that time, there weren't really any popular basic neuroscience YouTube channels. So my partner and I started talking about doing a series – we wondered what it might look like if we tried to create an "introduction to neuroscience" course using short videos and animations instead of hour-long lectures on a chalkboard.

We produced that first set of videos in the fall of 2015 and just kept going from there. We originally planned to just focus on basic topics, like "What is a neuron?" and "How do we see?" But eventually we started expanding into other topics that were fun, or culturally relevant, or that were requested by friends or family.

Almost 5 years later, we now have over a hundred videos on the channel and continue to produce content regularly. My partner has gotten freelancing filmmaking and editing gigs as a result of our work, and I spent 3 years freelancing for another YouTube channel, SciShow, in addition to writing for our channel.

Neuro Transmissions is something you've done while also working on your PhD. How did you balance the two?

It was a challenge, I won't lie. But I essentially saw Neuro Transmissions as my hobby; it's where I escaped from the lab and focused on doing something else that made me feel productive and creative. On top of that, I made a concentrated effort to maintain some semblance of work/life balance with my research.

So I spent my nights and weekends on Neuro Transmissions. I'd go to lab and work all day and then come home to make dinner and then spend a couple of hours writing scripts before bed. We'd film videos on weekends and post them on Sundays, so we could spend time that morning sharing the video around the Internet.

It kept me really busy, but I managed by creating balance in other ways like having a weekly game night with friends, so I always had some time to get out of the house and talk to other people and keep careful track of my calendar so I could get ahead when things were less busy.

I also learned how to say "no." There were so many different things I could have gone after in my research and lots of potential side projects – but I tried to keep my focus on my research goals and not let myself get distracted, so I could maintain that balance.

Saying "no" was important in other realms too. I eventually stopped my freelancing work so I could focus on my own projects and my research, and near the end of my PhD, while I was writing my dissertation, we stopped producing videos entirely for a few months.

It was hard to keep so many plates spinning, but I loved what Neuro Transmissions offered me – it was a place where I could make something I was proud of to share with the world, which was very helpful for coping with failed experiments and uncertain timelines. I found it very helpful to have some external validation when things in the lab were tough.

If someone was currently in their PhD studies, how might they decide if science communication might be a good fit for them?

All scientists have to do some science communication, even if they're not interested in doing outreach, so I think it's wise for most people working on their PhD to try and get some form of communications training. Even in a strictly academic career path, you need to write papers and grants and give talks to be successful, so working on those skills is beneficial.

If someone enjoys talking or writing about science as much or more than they enjoy doing the lab work, considering careers in scientific communications might be a good option for them. And careers can include highly technical writing (like papers and grants) or more public-facing communications (like writing news stories), so no matter how specific your interests, there are options out there.

If someone was interested in pursuing a career in science communication, what would you suggest they do to better prepare themselves?

I suggest that people start thinking about what they'd like to do next early on. Don't just assume that you'll want to do a postdoc or that you'll want to be a professor at the end of all of it; think hard about what that career path looks like and what you want out of your career.

Grad school is a lot of work, but your schedule will never be more flexible. So I recommend that students explore career paths – attend panels and networking events; apply to attend workshops; surf around on social media; just get an idea of what's out there. Try to connect with people at different career stages in areas you're interested in, either on social media or on LinkedIn or at conferences. Ideally, choose a PI who is understanding of the reality that not every grad student will go on to an academic career, and spend time talking to them about your interests and goals. Ask if they have any suggestions or know any they can connect you with in those arenas. Just like getting a postdoc often relies on having some personal connection or contact with the faculty in question, having a network of people in your chosen career path will help you find your next position.

I also suggest that people just start creating. Write blog posts; make videos; post pictures on Instagram – whatever kind of content you're most interested in producing, just start producing it. It will take time to build an audience and you'll make mistakes, and that's okay. When I go back to watch our earliest videos, I cringe because we've improved so much since then! You'd never expect to play at Carnegie Hall when you first pick up a violin, so don't feel like you need to instantly go viral to be a successful science communicator. Try stuff, see what you like, see what works for you, and then try again. Building a portfolio of communications work will also be beneficial for applying for jobs, to demonstrate that you can actually do the things that you're trying to market yourself for.

Consider that in any career path, you have to have some kind of personal brand. This doesn't mean you have to go hard on social media or use buzzy words; it just means that you need to be consciously thinking about how you're presenting yourself. Think about the kinds of jobs you want and think about the kind of person who would fit that position; present yourself professionally and thoughtfully online. Consider investing in a personal website to share your projects and make yourself easier to find. Having a digital presence makes it easier for potential employers to learn about you and what you do, and in a competitive job market, anything that makes you unique is valuable.

How similar or different do you think is your current role as science communications fellow from the work you've already been doing for Neuro Transmissions?

In both cases, I get to read about interesting research and often talk to scientists about that research and then come up with a way to share that story with whichever audience I'm hoping to connect with. It's dynamic and challenging, and I have a lot of freedom in figuring out what stories I want to tell and where they're best shared.

As a marketing and communications employee, my work is all in the name of the institution; I'm focused on sharing stories that align with the university's brand and that meet the goals of the department I'm working for. I only work with scientists in the university and sometimes their collaborators, and my writing is generally only published in university publications (like press releases, news stories, campus newsletters, podcasts, videos, etc.). When my work does get shared more broadly, it's usually anonymously, like in a press release, or working to connect scientists to outside media so they can tell their stories directly to the reporters.

At Neuro Transmissions, I have more freedom, but in some ways, there are more limitations. I'm responsible for all of my own brand management and coming up with all of my own ideas and stories. I have to do my own marketing and social media. I can make videos about anything, but sometimes, having so much freedom can be paralyzing, and all of those "extras" can be a lot of work! It can be nice having a team where jobs are clearly defined and you know exactly what you're responsible for.

What do you like most about your work?

I like getting to talk to all kinds of researchers about what they're working on and what they're hoping to accomplish and then translating that conversation into meaningful media to share with people outside of our institution. It's exciting to learn about new research and figure out why and how it's important and then to experience the challenge of figuring out the best way to share that story with the most relevant audiences.

I love the team that I work with at UC San Diego. I've gotten to work with several different media departments (almost like grad school rotations!), and it's been educational and rewarding to work with different teams and learn about their communications goals and team and management styles. The team in Health has been incredibly supportive and provided great mentorship; they've helped me refine my interests and learn new skills, which led directly to my next job offer, as a senior science writer at UChicago Medicine. I love that I get to talk to scientists all the time and learn about their research. I feel truly plugged into the pulse of what's happening at my university, and I love getting to help translate that to our community. I feel more connected to the city of San Diego now as a sci comm fellow than I did during all 6 years of grad school!

What do you like least about your work?

Dealing with people's egos. We have to be selective with what we share and want to be careful about how we're presenting information – sharing groundbreaking work without overpromising. But some researchers are convinced that their work is the most important, or that it's something that should be shared with the public even if it's not really going to resonate with our audiences, so I have to tread carefully. One reason I think it's good to have someone with a research background in this sort of position is that it also helps me push back when a researcher is making unsupported claims about their work – I can evaluate the study myself (to an extent) and can push back to make sure that what we're sharing is actually accurate. Researchers don't always love that, though – especially because they often don't realize I have a research background myself and make a lot of assumptions about my expertise, which can also be frustrating!

Sometimes, being limited by the branding guidelines and communications goals of the department can be challenging because I'm not used to being confined by those parameters. I'm used to getting to tell stories about whatever I want! I really think the key to enjoying branded communications is being part of a team that you feel good about and working at an institution whose values you agree with. Even when I feel a bit hindered in my storytelling, I still know that the work I'm doing supports an institution I care about and feel good about being part of. Also, it's hard when you're only spending a short amount of time with each team to figure out how to best make the transitions between them and navigate the new managers and team dynamics. I think this is probably true in many short-term fellowships and internships.

Based on your journey, what is some advice or suggestions you would want to pass on to someone who's currently finishing their PhD?

I recommend a lot of the same things that people would recommend for any non-academic career path, which can sound cliché but are so truly important. Like academia, getting a non-academic job requires knowing how to market yourself and knowing the right people to get you into the right places to share your knowledge. In academia, that means identifying mentors who can write good letters of recommendation and introduce you to potential postdoc advisors, writing grants, and attending conferences and workshops to network and share your research. In non-academic jobs, that means finding other ways to network and expand your skill set.

This might look like building a network on social media, including LinkedIn; tell people that you're interested in learning more about X career path, and as you make connections in that field, set up informational interviews. Look at job boards in that career path and see what kinds of skills and expertise are necessary, and figure out what kinds of courses or workshops you might be able to access to get you those necessary skills. And this is a big one – look for internships and/or fellowship opportunities in the path you're interested in! I'm 100% certain that my time as the Bigelow Science Communication Fellow was critical for landing me my next job because it was putting my money where my mouth was – I was actually doing the branded communications work that I said I wanted to do. It expanded my portfolio and gave me access to new resources and training. People interested in science communication careers should look for fellowships that align with their communications interests – the Bigelow Fellowship for those interested in PR/branded communications, an AAAS Mass Media Fellowship if you're interested in science journalism, or the AAAS Science and Technology Policy Fellowship if you're interested in policy communications. You can find these opportunities on social media, especially in places like the Science Communication Job Board on Facebook or the SciComm Board website (https://www.scicommboard.org/). And lots of groups and organizations are working on creating new opportunities in this arena, so keep your eyes open!

A lot of people like academia because they feel it gives them an opportunity to work on a topic that they deeply care about. Do you think this is also true in your current position?

I do! Because I'm part of a team, I'm not responsible for *all* of the communications efforts at my institution. That means that different people tend to have different beats based on their skills and expertise. Plus, part of what made this job exciting for me was knowing that it would give me the opportunity to continue to work on things I was really interested in – like making science videos and writing about neuroscience. But I also get to write about all kinds of things and work on a lot of different projects.

If someone is really interested in focusing on a particular topic in a science communication career, they may just need to be specific with the kinds of jobs they're applying for; if your interest is really in mental health, for example, apply for jobs with institutions or organizations focused on mental health research and resources. One reason I'm really excited for my next job is that I'll be the senior science writer, meaning that my beat will be biomedical research, with less of a focus on clinical care, which is less of an interest for me.

Another reason many like academia is that they feel it provides them with more freedom than they think they would get in other positions. How much freedom do you feel you have to work on what you think is interesting?

I think this is the sort of thing that will vary from place to place, but in general, I feel like I have a lot of freedom. Part of the motivation for the Fellowship is to create a space for a Fellow to produce new, unique stories that wouldn't otherwise be covered, so I've been heavily encouraged to seek out interesting ideas. I anticipate that I will still have a lot of freedom in my next job too – because even though an institution will have their specific communications and marketing goals, part of the job includes figuring out what kinds of stories will resonate with the audience. So I know that a large part of starting my new job will be getting to know the faculty on campus and learning about their work, and especially what kinds of exciting news they might have coming up, to think about the best ways to share that work with our audience. But ultimately, I am still beholden to our department's goals, and generally stories focus on new papers being published or culturally/socially relevant news topics.

I don't think it's as free or independent as grad school, but I actually don't mind that. For one thing, I personally find it helpful to work on a team where roles and responsibilities are clearly defined; it means the team is much more efficient and communicates more clearly about who needs to be doing what. It also gives me more space to manage my own work/life balance. If I know exactly what I need to be doing at work, I also know when I'm done, and I can shut off. That makes it easier to create space for Neuro Transmissions and for my other interests and hobbies. For me, it's worth sacrificing a bit of the freedom in my professional life to have more of that freedom in my personal life.

Is there anything else you'd like to tell someone reading this interview?

I guess the one big thing I'd say is even though I don't "need" my PhD for my current work, I don't regret going to grad school. I think we get used to this narrative that a career path is a straight line – grad school to postdoc to faculty position – and that used to be the case outside of academia too, when folks could get a good job right out of college and work there until their retirement.

The reality is we grow and change as we get older. What I wanted at 24 is not the same as what I want now in my early 30s – in my personal life or in my career. And that's okay. I have sometimes felt guilt and sadness over leaving research; I mean, I'm letting go of something I dreamed of for years and spent a long time working toward. And in America, at least, grad school really sets you up to make being a graduate student a part of your identity. It's okay to grieve the loss of those things. I had to make a leap out of the standard path and figure out something new, and I wasn't totally sure it was the right choice. But now, almost a year after defending, I feel so good about where I am and where my decisions have gotten me. I just had to learn to accept that even though I have a lot of interests, there are only so many hours in the day, and I am just one person. Sometimes, you have to sit down and really break down your priorities – and sometimes, you might realize that the things you thought were the most important actually aren't.

And what if I had been wrong? What if I'd left research and discovered that I truly did need to be at the bench to be happy? There's no law saying I couldn't go back. I could still leverage the connections I have in academia to find a postdoc, or try for an industry position, and get back into the lab. Your career doesn't have to be your identity, and trying something new isn't a death sentence. So explore. Try things out. And when something makes you happy, keep doing it!

Thank you so much for sharing your experiences with us! I'm sure your thoughts here will help many as they find their own career paths.

Chris: It's been several months since your interview and you have had more experience in your current position now, can you tell us more about your current job?

Alie: Absolutely! I'm currently working as the "senior science writer" at the University of Chicago Medicine, where I essentially act as a scientific public information officer. I work with faculty in the medical center and the Biological Sciences Division to share their research with the public, which includes writing press releases and news features, putting out research newsletters, sharing news on social media, developing multimedia projects, and acting as a liaison with traditional and online media outlets. Lately, that work has included a lot of communications around COVID-19 research, treatments, and vaccination. I feel very lucky to be in a position where I can work 100% from home, and in fact I believe my position will remain primarily remote even after the pandemic ends, which I don't mind. Writing press releases in slippers with a cup of tea and a cat on my lap isn't too shabby. It's been very difficult starting and adjusting to a new position during a pandemic, but we're all figuring it out as we go. I've been in my role for 6 months now, and one thing I really appreciate about my position is that I can see room for it to grow. My team is very supportive of helping employees develop new ideas, keeping them challenged and fulfilled, for the benefit of our department. So in addition to my writing work, I'm starting to develop some new plans for identifying and reaching key research communications goals, as well as developing new resources for scientists to support their own communications interests and efforts. I'm still learning about all of the faculty; it's a lot harder to know who's who when you can't just go visit people in their lab! But it's great to be at such a stellar research institution in such a vibrant city, and I'm grateful to have this opportunity. I also love that the work compliments and supports my own science communication interests; I use a lot of the same muscles for my day job as I use when working on Neuro Transmissions, and it's nice to be able to expand my skills in both arenas.

I think the pandemic really upended things for everyone, and I have definitely caught myself starting to worry about "what's next," but I've recently decided that I need to just settle down and focus on what I'm doing for a little while. There will be plenty of time to figure out what's next, and in the meantime, I'm grateful to have a job that I enjoy, on a team that wants to help me grow, in a city that I'm already growing to love!

It's great to hear that you're doing well. Thank you for the update!

“I Am Pretty Interested in Coding, Technology, and Infrastructure”

Arfon M. Smith

Arfon M. Smith

Abstract In our interview with Arfon Smith, we discuss how an interest in writing research software can lead to positions with organizations such as Zooniverse and GitHub. Though not yet established at the time, this type of position is characterized by the term ‘research software engineer’. Arfon himself has helped pave the way for this career path, through his efforts in founding the *Journal of Open Source Software*. The traditional academic path is not a good fit for everyone, but there are many shared skills involved in PhD research and working as a product manager.

Contents

A. M. Smith (✉)
Space Telescope Science Institute, Baltimore, MD, USA

GitHub Inc., Edinburgh, UK

C. R. Madan (ed.), *Academia and the World Beyond*,
https://doi.org/10.1007/978-3-030-82606-2_10

Chris: Can you introduce yourself and tell me a bit about your current position?

Arfon: My name is Arfon Smith. I work at Space Telescope Science Institute (STScI) in Baltimore, USA. STScI was founded in 1981 to run the science operations of the Hubble Space Telescope. Since 1990, STScI has been the operational interface for Hubble, serving the global astronomical community who make use of this flagship facility. A key part of our work as the *science operations centre* for Hubble and the soon to be launched James Webb Space Telescope (JWST) is to ensure the continued scientific legacy of the missions. Capturing and preserving the data associated with Hubble is the responsibility of the Barbara A. Mikulski Archive for Space Telescopes (MAST), which is the archive for Hubble and more than 20 other mission datasets including Kepler, TESS, IUE and Galex. MAST currently holds data from 21 missions and surveys and, with a data volume of over 2 petabytes, is a major infrastructure support effort in and of itself.

As the head of the Data Science Mission Office at STScI, I am responsible for defining the organization's strategy and roadmap for data management and data science with our multi-mission, multi-petabyte archives, and I am a member of the senior leadership team of the institute (~750 staff). My team manages a portfolio of work for approximately 100 engineers, astronomers and data scientists.

What was the focus of your PhD?

My PhD is from the University of Nottingham (2006) in astrochemistry (strictly speaking just 'chemistry'). I was situated in the chemistry department at Nottingham. The focus of my studies was to probe the small-scale structure of the interstellar medium in our galaxy (the gas and dust between stars) using spectrographs attached to telescopes.

As you were finishing your PhD, what were you thinking about your career plans?

I wasn't too sure to be honest. I had a pretty strong sense that a traditional academic career really wasn't going to work out for me. Clues/signals that led me to this conclusion included me not really being that interested in keeping up with the literature in my field, being more interested in the software I was writing than the results/insights I was producing! I applied for a couple of postdocs because I couldn't really think of anything to do but wasn't successful.

I then embarked upon what became a career building infrastructure/writing software to support academic research (Wellcome Trust Sanger Institute, Zooniverse, GitHub, STScI).

Can you tell us a bit about what day-to-day life is like in your current position?

My current position is pretty mixed. My 'day job' (and roughly 80% of my time) is in a leadership role at STScI. The small team I lead are effectively 'portfolio managers' in the sense that we coordinate a portfolio of work for a small collection of other organizational units (e.g. Hubble Space Telescope (HST), James Webb Space Telescope (JWST), Wide Field Infrared Survey Telescope (WFIRST)). I would say my job roughly breaks down into the following four categories:

1) Spending time with other leaders making and shaping decisions about priorities for the organization, putting together organization-level strategies for data management and data science and developing roadmaps/priorities for individual teams.
2) Working with individual product teams to give feedback on their progress executing on roadmaps.
3) Developing new ideas for projects and exploratory R & D with new technologies and product ideas.
4) Reporting on work: As a government contractor, we have a lot of reporting to do to the government (NASA in our case).

The final category of work is the other 20% of my time. I am fortunate to have 20% research time which I almost exclusively use to support the *Journal of Open Source Software* (JOSS, https://joss.theoj.org) as editor-in-chief.

You've had a few positions between completing your PhD and your current position, can you tell us a bit more about how these positions differed?

I had about 3 years in what I would now call a Research Software Engineer position although that title didn't exist at the time. Firstly, at the Wellcome Trust Sanger Institute in Cambridge building tools to support DNA sequencing experiments. I only held this position for a year before joining the Galaxy Zoo team at the University of Oxford as a postdoc, spending all of my time doing engineering work. Ultimately this led to the development of the Zooniverse platform (https://www.zooniverse.org) which I co-founded and spent the next 3 years leading from a technical standpoint.

I then moved out to the USA to lead a new grant designed to grow the Zooniverse into a true platform for global citizen science. This grant was hosted at the Adler Planetarium, and so I essentially held a dual role here – Director of Citizen Science (at Adler) and Technical Lead of the Zooniverse. This was my first real experience of managing people (I had a team of 12 people at Adler) and leadership (as a member of the management team).

In late 2013, I joined GitHub to lead their engagement with the scientific research community. With a focus on open source software and open data in science, my responsibilities included working with industry partners including journals, data publishers and funding agencies to develop strategies for sharing data and software in academia and improving the GitHub service offerings for academic users. Examples of projects I worked on include Zenodo-GitHub integration (https://guides.github.com/activities/citable-code/), Jupyter Notebook rendering (https://help.github.com/en/articles/working-with-jupyter-notebook-files-on-github) and Software Citation Principles (https://peerj.com/articles/cs-86/).

Later I transitioned to Program Manager for Open Source Data, and in this role, my focus was on ensuring that GitHub remained a responsible host of open source communities and their data and that GitHub maximized opportunities to develop a deeper intellectual understanding of how open source communities work. My core areas of focus were:

1) Working with our platform team to improve open source data products produced by GitHub
2) Facilitating third-party research around GitHub data products and network activity
3) Working with open source communities, enterprises and research groups to develop industry standard metrics for measuring open source health
4) Leading the recruitment of GitHub's first data science team

If someone currently finishing their PhD was considering a position similar to yours, how might they decide if it would be a good fit?

The common thread in my career has been designing and building tools that support the work of academics and researchers. Even though I came late to software (I wasn't into computers growing up, for example), I am pretty interested in coding, technology and infrastructure. I also really like *building things* (I always have a side project on the go, for example).

I think the closest industry term for the sort of thing I've been doing most of my career is *product management* (with my STScI gig being one level more abstracted away as a *portfolio manager*). There's a bunch of articles out there about what it looks like to be a product manager (e.g. https://hbr.org/2017/12/what-it-takes-to-become-a-great-product-manager, http://blogs.nature.com/naturejobs/2018/06/06/how-product-management-could-be-a-route-out-of-academia-for-phds-and-postdocs/). I know a fair number of people ex-academics that have transitioned to product management and are very happy.

If someone was interested in pursuing a similar career path, what would you suggest they do to better prepare themselves?

I think my interest in technology and software has stood me in pretty good stead career-wise. My current role and in part my previous ones too require me to be pretty current with technology trends in both the current research discipline I'm actively supporting (astronomy/astrophysics) and in the wider economy. While this isn't true of my current role, it was at GitHub so I think it's worth sharing here: many companies care much more about your skills and ability to integrate into their workforce than they care about how many papers you've published! So if you do find yourself moving out of academia and into industry, then some skills are much more valuable than others. For example, in an engineering or product company, knowing about Git(Hub), Python and backlogs/Kanban/Scrum/how engineering works is managed – all of these things are much more important than whether you have one, two or three papers in that prestigious journal your field cares about.

While I'm on the topic of transitioning out of academia, I think it's really important that you substantially rewrite your resume if you are looking for positions in industry. Ask someone you already know who has made the transition to help you with this. They should be able to help you reframe your knowledge/skills/expertise for a non-academic audience.

A lot of people like academia because they feel it gives them an opportunity to work on a topic that they deeply care about. Do you think this is also true in your current position?

I do yes, but I think I'm only in this position because of a couple of factors: (1) being willing to move around between jobs semi-regularly and having a family able to support this and (2) some luck (right job, right time opportunities, etc.) There have been a number of times in my career when I had a 'solid' job that there was no good reason to leave (but I did). Each time I've been pleased with that change, but there's been some risk too I think.

Another reason many like academia is that they feel it provides them with more freedom than they think they would get in other positions. How much freedom do you feel you have to work on what you think is interesting?

Hrm, I actually think this is a fallacy for many in academia. You've accidentally hit on a topic I promised myself I'd write a blog post on a few years ago and then failed to (!) so I'll attempt to summarize here....

On the face of it, I agree, academia has lots more freedom. That said, in my experience, many academics (and more importantly their projects) are severely constrained by the funding landscape they exist in. As a result, it's a pretty common game to try and fit your ideas to some kind of funding opportunity, that is, you end up having to *write to the grant*. An extension of this would be actually selecting a potential research direction *because* of the funding opportunities available to them. In the worst-case scenario, your research agenda is effectively being directed or at least heavily influenced by the interests/preferences/whims of your funders. This is obviously an extreme I'm describing here, but I believe I've seen this in action multiple times in a variety of academic settings. So in summary, in the idealized form, academia has lots of freedom, but in reality, I don't think it's quite that 'pure' or straightforward.

In industry, there can be lots of freedom to work on new ideas too, provided the business is doing well and there's a focus on innovation. When a business decides to take on a new challenge, there are often substantially more resources to apply to a problem than in academia. In my experience of industry work and projects, it's often easier to stay true to your ideas and more intellectually honest about the work you want to do. Easier to 'stay true' because you're less resource constrained, and money often doesn't have many restrictions. More intellectually honest because if you've made a pitch for a new product/feature/line of business and customers hate it, then there's a good chance it's not as good an idea as you initially thought....

What do you like most about your work?

I really like working for an organization that has a large impact on the community. For example, because of the nature and scale of the science missions that STScI operates, a significant fraction of the available funds for open source software in astronomy are spent by us. This means that decisions we make as an organization can affect real change on the global astrophysics community. We take this responsibility very seriously, and I'm fortunate to be part of an incredible team of astronomers and engineers responsible for making these decisions.

And what do you like least about your work?

The downside of being at STScI is the overhead of working with the government which is substantial and affects work in interesting and surprising ways. NASA is an amazing organization that has done and continues to do incredibly ambitious things that further our scientific understanding of the universe, but fundamentally, working within government bureaucracy can be tiring and very frustrating at times.

Based on your journey, what advice or suggestions do you want to pass on to someone who's currently finishing their PhD?

If you can, talk regularly to people that know you and whose opinion you trust. One thing I've benefited from immensely is having a small number of people who I've been able to bounce ideas off about possible career moves and professional opportunities. The most important of these people for me is my wonderful partner Laura – her counsel, support and willingness to embark on crazy adventures have been an essential part of my professional journey.

Is there anything else you'd like to tell someone reading this interview?

I consider it part of my job in life to try and help others with their careers, especially people who are thinking that a traditional academic career path might not be right for them. If you want to talk more, feel free to hit me up on Twitter (@arfon).

Thank you so much for telling us about your work, Arfon!

Chris: It's been a few months since your interview and you are in a new position now, can you tell us about your new job?

Arfon: Sure thing! I've actually returned to work at GitHub in the product organization (I previously worked at GitHub from 2013 to 2016). There was absolutely nothing wrong with my job at STScI (I *really* enjoyed it), but ultimately my family and I decided we needed to be back in Europe to be closer to family.

In my new role, I'm a product manager in the Data organization at GitHub. Data includes teams responsible for data engineering, data science/machine learning/AI, analytics and more. The work of these teams is realized in a number of ways including working with product teams not only to build new customer-facing functionality (e.g. data-powered products) but also to support internal teams who want to understand the business through insights derived from data.

As a product manager, my role is to guide the work of these teams, helping us all understand who our customers are, what the most impactful work is that we should be doing right now. Working with leadership to set a vision for the Data organization for the next 12–24 months is also an important part of the role.

Can you tell us more about how your previous GitHub position differs from your new position?

In my previous position, I was in more of an outreach/engagement role with some program management. Ultimately this work was quite mixed – connecting lots of threads across the organization, making sure that GitHub was supporting the needs of particular communities either working on or researching the platform.

You can find lots of articles about the difference between program and product managers (e.g. https://medium.com/pm101/the-difference-between-product-program-and-project-management-64e2f1ee4f01). In truth, there's lots of overlap, and depending upon the organization, I think they can be very similar. In my case though, the biggest difference is that I'm now directly working with a number of engineering and data science teams, which means my input and direction directly influence the ongoing work of these teams. In my previous role, I had very few engineering resources allocated my way so had to spend much more time advocating for a particular approach/direction.

In short, I think my new gig is an upgrade on the last one at GitHub.

How do you think having a PhD helps you succeed in your current position?

Yes, I think it does. There's a fair amount written online, mostly in blog format or presentations, about how academics can make good product managers (e.g. http://blogs.nature.com/naturejobs/2018/06/06/how-product-management-could-be-a-route-out-of-academia-for-phds-and-postdocs/). Some of the skills cited include communicating ideas effectively, working with broad cross-functional teams, having focus in a position of uncertainty and knowing how to ask the right questions and use data to help make decisions. All of these are traits that are generally desirable for product managers but are also core skills for research. While I would never tell anyone to go get a PhD in order to become a product manager, I think there's a surprising amount of overlap in the skills required to be successful across the two roles.

That was very informative. Thank you for the follow-up interview!

"There Is an Enormous Market for PhDs in Technical Sales Positions"

Cleyde V. Helena

Cleyde V. Helena

Abstract In our interview with Cleyde Helena, she tells us about her research work and how lack of continued funding led to the lab she was working in to close. Cleyde now works as an independent contractor in technical sales, helping researchers purchase customized scientific equipment for their research. Leaving academia was a long and painful process for Cleyde, but now she helps others realize that other career paths can also be fulfilling. If you are having doubts or are feeling pressured by others, think about why you want to stay or leave academia. There are many non-academic careers out there; it is worth considering if they might be a better fit and make you happy.

Contents

C. V. Helena (✉)
ASC Group LLC, Centerton, AR, USA

C. R. Madan (ed.), *Academia and the World Beyond*,
https://doi.org/10.1007/978-3-030-82606-2_11

Chris: Can you introduce yourself and tell me a bit about your current position?

Cleyde: My name is Cleyde Helena and I am a Brazilian-Chilean-American living in the United States. I am part of an analytical chemistry consulting group, and I represent four different companies. My official title is 'account manager', but I feel like 'problem solver' would be more appropriate. I am the person that links the customer to all parts of the organization, and although my salary is a commission on individual sales, I spend most of my time dealing with mundane tasks: emails, virtual meetings, customer training and phone calls.

I sell specific products within a geographic region. Because of my background in academia, they gave me an academic territory, so I cover all academic and non-profit institutions located in this territory. I enjoy this because it allows me to continue in the academic environment but without the pressure and stress that I was used to. Even though most of my work can be done remotely, I occasionally travel to meet customers face to face. Most of the instruments that I sell are very expensive, and it is usually very beneficial to meet in person. First to establish a relationship of trust but also to make sure you understand all requirements needed and suggest the optimal instrument/configuration for their research.

I am not a chemist, and it still takes me some effort and research to fully understand the several different types of chemical analysis that come to my hands. Also, I work within a team, so I can easily ask the product specialists for advice about application and instrument configurations. Most of the professors have an idea of what they want to purchase, but it is my job to make sure that the instrument is fully equipped to perform the analysis.

What was the focus of your PhD?

I have a PhD in Human Physiology, with focus in neuroendocrinology from the Physiology department at the School of Medicine of the University of Sao Paulo, Brazil. Our lab studied the neural control of female endocrinology. In short, how the brain controls the hormonal secretion in females. Our lab was particularly interested in the role of norepinephrine (particularly the one coming from the Locus Coeruleus) in mediating the steroid-induced gonadotropin surges that lead to ovulation.

In my master's, I studied the effect of Locus Coeruleus lesions on the preovulatory surges of LH (luteinizing hormone) and FSH (follicle-stimulating hormone), as well as the one induced by ovarian steroids in ovariectomized rats. In my PhD, I investigated the specific role of estradiol and progesterone in the Locus Coeruleus under several steroidal conditions. Interestingly, I've got in touch with a professor at the Rockefeller University, and when I attended my first Society for Neuroscience meeting (back in 2002), I also took the opportunity to visit their lab in NYC. It was my first time visiting the United States, and I fell in love with … everything! We decided to apply for a Sandwich fellowship in Brazil, so I could come to their lab and do part of my PhD research in the United States. After spending 1 year doing research in an American lab, I was certain that I wanted to come back and pursue an academic career in the United States.

Because of visa restrictions, I couldn't come back to the United States right away after defending my PhD. I decided to do a postdoc in the same lab where I did my PhD because it was the easier thing to do, and there weren't a lot of people in Brazil working with neuroendocrinology of female reproduction. Looking back now, I believe this was one of my initial 'mistakes'. On one hand, continuing in the same lab gives you the opportunity to continue and finish several projects you've been working on. But, on the other hand, it doesn't give you the opportunity to grow, to learn new techniques and to pursue new ideas.

After 3 years, I was invited to apply for a postdoc position at Florida State University. It was also to study the neural control of female reproduction but focusing on the dopaminergic control of prolactin secretion. I was so excited about finally coming back to the United States and pursuing my academic dream. This new lab had a somewhat hybrid group, with part of scientists working on in vivo projects and others performing in vitro projects, and all data would be combined and modelled by some computational neuroscientists. This new experience was challenging and intimidating at first, but it turned out to be a valuable growing experience. A mathematician's mind functions in a very different way than ours, biologists. We tend to complicate things, to be very verbose. They are incredibly sharper; they go straight to the point. Working directly with them taught me to be more direct and concise. Sometimes, I think that they simplify too many things, but in the end, it's like they are summarizing things in a box and arrow diagram by the end of our paper.

As you were finishing your PhD, what were you thinking about your career plans?

Some people have traumatic experiences during graduate school. I didn't. Of course, graduate school was hard, and it took me years to finally really understand what I was doing! But when I was finishing my PhD, I finally started to see the big picture of my research, and I was really excited to continue doing it. I've already started attending international congresses, and meeting those famous researchers in person only made me aspire to be like one of them someday. I'd dream about migrating to the United States and getting an academic position in a nice university. I had no idea back then of how hard that would be!

Can you tell us a bit about what day-to-day life is like in your current position?

Since I've transitioned out of academia, I've been working from home. I spend most of my time on the phone with customers and coworkers, responding to emails, creating and editing quotes. As an independent contractor, I don't have a salary and work on commission. That can be quite stressful as academic sales depend on grant funding and so many other things! So the main priority of my daily work is to generate sales. The analytical equipment that I sell are complex solutions, and you cannot get any pricing on our websites. Every instrument is customized, and every sale requires discussions about what each customer needs. As I've been in this job for a while, I usually can configure most of the instruments. However, sometimes, there are more complex solutions that require asking for help from our product specialists or other specialized people. Although I am not responsible for handling post-sales installations and service, I am usually contacted or copied in these emails as well. As an account manager, I am expected to be the point of contact between customers and the company. Even though nine out of ten times I cannot personally help the customer, I certainly know who can!

I also try to make regular visits to the several campuses I represent. The instruments I sell are very expensive, and it is useful to meet professors in person to create a relationship of trust. When meeting in person is not possible, we perform online meetings and presentations. We are also expected to organize at least two local seminars per year, on a topic that would be of interest to a large number of local researchers. Organizing these types of events is usually a lot of work but also very fulfilling.

Technology is constantly changing, and there are always new techniques, new instruments, new challenges. Because of that, I normally attend at least one online training per week and receive lots of emails with material to read on your own time. I thought it would be harder to work from home and deal with the procrastination and distractions. But when you deal with people, everybody has their own needs, and most of the things are urgent. When I was in academia, I used to have a to-do list and normally complete each task in the order of my list. Currently, things get juggled to the top in order of urgency. Most of the tasks take days, sometimes weeks, to be finished, making it harder to keep track of everything! The good thing is that there is no shortage of work and there is no day like another. Every day is new and different, and that's one of the reasons I really like my job.

How do you think having a PhD has helped you succeed in your current position?

There is no doubt that having a PhD has helped me succeed in my current position. The main reason I was hired was because I had a PhD and have spent many years in academia. That helps me to understand the professor's needs and struggles, making it easier to communicate to them. Because I am not a chemist and don't have much lab experience with the equipment I sell, I need to read a lot of literature. Before talking to a new professor, I always go to their webpage and read at least the abstracts of their latest publications. Sometimes, professors send us some literature that performs the analysis they are willing to set up in their laboratory, and I can see that most of my non-academic coworkers either don't understand or don't have the patience to read scientific literature. Also, the presentation skills I've learned during my PhD are very helpful when putting a PowerPoint together to present to a customer. I attend training talks with PowerPoint presentations on a weekly basis. It is so easy to identify which ones were prepared by a PhD! Even upper management people with lots of experience do not present their talks in an organized and storytelling manner as we do. Other very useful skills I use on a daily basis are time management, multi-tasking and ability to work under pressure and deadlines.

If someone currently finishing their PhD was considering a similar position as you have now, how might they decide if it would be a good fit?

Some people say that you need to be an extroverted person to succeed in sales. Being extroverted certainly helps, but the number one point in my opinion is that you have to be a people person. On average, I probably spend over 4 h/day talking to people either on the phone or in person. So not only you should enjoy talking to people but also understand them. Read between the lines and know when you need to explain all the details about a certain aspect, or when to only focus on the main parts. You need a lot of patience and enjoy helping others. An equipment sales person needs to be able to start and maintain relationships. The sales cycle is usually long (typically about 1 year or more), so you need to find creative ways to keep in touch with your customers during the waiting periods of time. Follow up, but never be pushy. Being trustworthy is key. Each sale is based on trust, and you are not just selling equipment. You are selling a solution, and your role is more of a consultant than of a sales person.

If someone was interested in pursuing a similar career path, what would you suggest they do to better prepare themselves?

There is an enormous market for PhDs in technical sales positions, but not that many at the moment simply because most PhDs do not apply for those positions. That usually happens mainly because PhDs think they need to have sales experience to apply for these positions, but also because they think that sales people need to be pushy and annoying. That couldn't be further from the truth in technical sales! In this line of work, you are not usually selling a product, but you are selling a solution. Usually, the customer has a deficiency that needs to be eased, and you are the consultant that is going to guide them through the process.

If you'd like to pursue a career in technical sales, I would suggest having informational interviews with people currently working in the technical selling branch you would like to explore. That helped me ease the stigma associated with sales people! That would also help you visualize yourself into that future and realize if this feels right for you. Regardless of the career path you want to pursue, I recommend doing as many informational interviews as possible. You will be surprised at how willing people are to help and talk about their experiences and career. And most importantly, while it may not help you choose the path you want to take, it certainly will make you certain of the ones you definitely don't want to take.

A lot of people like academia because they feel it gives them an opportunity to work on a topic that they deeply care about. Do you think this is also true in your current position?

I feel very fortunate in my current position now. For many years, I've thought I'd never find happiness and fulfilment in any occupation other than academia. Right now I feel I'm in an academic-adjacent position. I am not actively performing research, but I am directly supporting professors to achieve their projects by providing them with adequate instrumentation. I am still directly immersed in the academic environment, constantly talking to professors and students, visiting campuses and aware of the academic calendar of classes and grant proposals. My field of research was neuroscience, and I had no idea of how many interesting projects are done in chemistry and engineering! I don't have the gratification of seeing my own papers being published, but every time that I see a professor publishing something achieved using the instrumentation I've provided, it gives me a sense of gratification. It is not the same proud feeling of seeing your own work being published, but still feels like you helped it happen.

Another reason many like academia is that they feel it provides them with more freedom than they think they would get in other positions. How much freedom do you feel you have to work on what you think is interesting?

For many years, I also thought that academia provides researchers with more freedom than in other positions. However, during my last years in academia, I started realizing that is not entirely true. There is a false sense of freedom that states that you can research whatever you want, at the pace you want. In reality, there are research topics that are more recognized than others. In other words, if you don't do research in a highly applicable/translatable topic, you might not be able to fund your research. Also, you need to publish at a consistent frequency. Not too little, not too much. It is okay to increase the number of publications, but it has to be at a consistent rate. You get rewarded by the number of publications, by the impact factors of the journals you publish, and most of the time early career researchers are forced to teach topics that they are not familiar with and to join committees that they are not interested in.

I personally feel I have way more freedom now. As an independent contractor, I am my own boss. Even though I report to a manager and technically have to follow rules and reach a sales quota, I do not have to report my daily work or have someone telling me what to do. As I work on commission only, if I don't sell, I don't have a salary. So it is in my best interest to sell as many instruments as I can, but the way I decide to do that is totally up to me. The good part of it is that I have the freedom to take several days off work when I want. The downside of it is that I don't have paid vacations, so I must plan accordingly.

Even though I still occasionally work over weekends and holidays, this is much rarer now than when I was in academia. During the first month or so outside of academia, I used to feel very guilty about not working on the weekends. I remember once my manager asked me why I was still working, as it was already 5:30 pm. It is incredible how this culture of overworking is common sense in academia, and you just realize how unhealthy it is after you leave academia.

Can you tell us a bit more about how you transitioned to a job adjacent to academia? In particular, apart from thinking about translatable skills and finding a suitable job, how was it to change your perspective to mentally transition to a position outside of academia?

Leaving academia was a long and painful process for me. I don't know exactly when I started to change my mind about leaving academia. I think it was a mix of things that led me out of it. There was a lot of frustration after tons of academic job applications (which require a lot of work!) and interviews that didn't turn into a job offer. On Twitter, I could see how almost everybody seems to be struggling to succeed, writing several grants every grant cycle with little success rate. I could see how doing research was getting harder and harder, especially in my field. I used to attend the Society for Neuroscience meeting, and I could see the transition happening over the years. People slowly leaving the field or applying for grants to study some other topics, and doing neuroendocrinology research 'on the side'. I remember talking to a researcher that changed fields because 'you gotta go where the money is'. True, but still depressing.

I still had a romantic view of science, of getting excited with data, with results. Not really worried about performing top techniques or publishing in top journals. I still wanted to do my basic research and not worry about including fancy techniques just because it would look good on my grant application. But then I started to believe I had to rethink my future research, not because I cared about those hot topics or I think they were necessary, but because I needed to get a job and money for my research. It made me feel like a research whore.

I believe the final disillusionment with academia happened when we received the news that our lab grant was not going to be renewed and our lab was going to close its doors. I am someone that always thought I had very clear goals in my mind. And for many years, I truly believed that academia was the only path that I wanted to pursue. The idea of becoming unemployed scared me, but not having a clear future goal scared me more.

After working in academia for almost 20 years, the hardest challenge for me was to find out which path I could pursue that I would be qualified for and that would bring me personal fulfilment. Because I've been on the academic one for so long, I had a hard time thinking about other paths. Also, not everything that you enjoy you will be able to do it effectively, and not everything you might be good at will necessarily be a gratifying job. It takes a lot of research and self-examination to find the best decision.

Leaving academia in many fields is the ultimate taboo. In journal clubs, seminars and laboratories, jokes are made that scientists who left academia for industry or other jobs have 'turned to the Dark Side'. A culture persists that academia is a noble calling that trainees should single-mindedly pursue. Academia is sticky, and when you are in it, everything funnels to pursuing an academic faculty appointment.

Only 12% of all PhD holders have tenure or a tenure-track appointment at an academic institution (see https://nsf.gov/statistics/2016/nsb20161/#/data, Table 3-16). Despite this fact, pursuing a career in anything other than academic research is still seen as 'alternative'. Worse still, the decision to leave academia is usually accompanied by feelings of failure. That prompted us to create the *Recovering Academic* podcast (http://recoveringacademic.net). Although we found articles and online resources about how and why to make the transition, we felt like there was a lack of emotional and community support to those that want to leave or are thinking about leaving academia.

Leaving academia, and re-charting a career path, can feel jarring. The first step, and possibly the hardest, is to make the decision to leave. Most PhD holders spend at least a decade working in academic environments. Moving into something unknown may seem like a rogue move. Most advisers and institutions do a poor job promoting and preparing PhDs for non-academic, so-called alternative careers. The academic environment trains PhD holders to learn how to do experiments and write manuscripts, but not how these skills translate to the 'real world'. PhD holders tend to think that they will leave behind all the skills they acquired while in academia, and they will have to start from scratch. But universities are not the only place where you can apply those skills! In fact, PhD level training—problem solving and project management—is applicable to most jobs. This may mean starting as a beginner in a new profession and rising quickly because of their doctoral training. Being in the ivory tower for so long, academics can easily fall into cognitive biases and forget that less than 2% of the population have a PhD. You might need to do some soul-searching, writing down a list of ideal career characteristics, for instance, to figure out what career path to pursue and how to transfer your academic skills towards your new role.

I got my first job outside academia in July 2015. Life changes and slowly I started to realize that academia was not the only way to be happy. Today, it's been over 6 years since I left academia, and I am more certain than ever that this was the best decision I could have taken. I admire those who continue in academia, but the more time passes, the more I feel like this was not the life I wanted for me. I feel like now I have a much healthier work-life balance and also a much better salary!

Based on your journey, what advice or suggestions do you want to pass on to someone who's currently finishing their PhD?

When we are working on our PhD, we spend so many years putting a lot of effort into developing our research and writing our thesis that it is easy to lose track of the big picture. Of course, we all want to finish our PhD and defend our dissertation, but what's next? It is beneficial to have a long-term plan in mind and have a list of goals that can help you achieve them, such as 'publish X articles/year'. Your long-term plan may change along the years, so your immediate goals need to be adjusted accordingly. It is harder than you think to figure out what would make something your 'dream job'. Don't be afraid to experiment and to discuss this with other people.

That brings me to the second suggestion: no matter what is your long-term goal, you need to focus on networking. It is not just about growing the number of LinkedIn connections or simply knowing a lot of people. It is about having some meaningful connections with whom you can exchange ideas and ask for advice. The more you talk to several people, the more you understand the options that are out there and that can bring a lot of new opportunities. A lot of times we want to find 'the path' that will lead us to our perfect career choice. But there isn't only one perfect path, there are many different paths to be explored, and many can indeed lead you to your ideal career. And if the path you are currently pursuing ends up not being the one that will make you happy, don't be afraid to stop, rethink and adjust your path.

Is there anything else you'd like to tell someone reading this interview?

Leaving academia is not easy. It takes a lot of courage to take the leap and jump out there in the open, in the unknown. Because we have been in academia for so long, we are surrounded by other academics, and it is easy to get swamped in feelings that if everyone around you succeeds in academia, you should succeed as well. And if you don't, you are a failure. This could not be farthest from the truth. Academia is not for everyone, and there are so many other career paths where you can use your PhD, directly related to higher education or not.

If you are having doubts about if academia is right for you, it is worth taking a moment to think and realize if you should keep pushing forward or start considering other options. If possible, you may want to take a break to see what is 'out there', but unfortunately, this is not an easy alternative to take. There are many reasons *why* you may be thinking about leaving academia. Is it because of the terrible job market, because of the lack of work-life balance or maybe because you simply realized that the academic life is not for you.

Regardless of your reasons, it is important to do some soul-searching and understand your own reasons before sharing them with the people that care about you. It is also very helpful to talk with other PhDs that have left academia and ask about their experience during the transition and their current state of mind. It is so easy to think that you're all alone and the only person going through those feelings. You will probably find out very quickly that most of the PhDs that left academia shared similar feelings and doubts about their choices.

When you decide to leave, there might be some pressure from immediate people like your advisor or family. Most advisors put a lot of effort into your academic growth and may see you as an 'heir' to their topic of research. Therefore, the idea of seeing a student leaving the path they imagined for you can be very disappointing. Immediate family may contest why you are willing to leave something that you, and them, in a way, have put so much effort and time pursuing.

There are plenty of opportunities for PhDs outside academia. You just need to make sure to find out which one will bring you fulfilment and make you happy!

Thank you so much for sharing your experiences with us, Cleyde! I am sure your advice and perspective here will help many.

"I Had No Idea That the Option to Pursue a Career as a Professional Scientific Editor Even Existed"

Stavroula Kousta

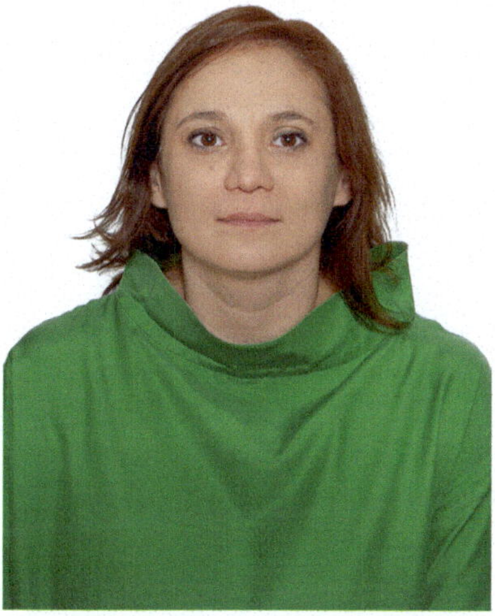

Stavroula Kousta

Abstract In our interview with Stavroula Kousta, we learn about her current position as chief editor for *Nature Human Behaviour*. This was preceded by several earlier full-time editor positions and can be a great option for those who love to read lots of papers and become involved in journal policy and may not want to focus on specific research questions themselves. As an editor, however, you can still champion topics that are close to your heart and help shape science and its communication. Stavroula suggests that PhD students consider their skill set and what they would enjoy doing when exploring different career paths.

Contents

S. Kousta (✉)
Nature Human Behaviour, Nature Portfolio, Springer Nature, London, UK
e-mail: stavroula.kousta@nature.com

C. R. Madan (ed.), *Academia and the World Beyond*,
https://doi.org/10.1007/978-3-030-82606-2_12

Chris: Can you introduce yourself and tell me a bit about your current position?

Stavroula: I am the chief editor of *Nature Human Behaviour* (https://www.nature.com/nathumbehav/), a multidisciplinary journal that publishes research of outstanding significance on any aspect of individual and collective human behavior. I started my career as a professional scientific editor back in 2008, when I became the editor of *Trends in Cognitive Sciences* (https://www.cell.com/trends/cognitive-sciences/home), a monthly reviews journal. In 2013, I joined *PLOS Biology* (https://journals.plos.org/plosbiology/), a peer-reviewed scientific journal covering all aspects of biology, as a senior editor. I was hired to launch *Nature Human Behaviour* in 2016 and have been its chief editor since then.

In my role, I am responsible for the journal's overall strategy and editorial policies. I oversee a small team of six editors, and together we work on the evaluation, peer review, and editorial decisions for research manuscripts submitted to the journal. Unlike the majority of research journals, we do not have an external editorial board – all editorial decisions are made by our team of in-house professional editors. We also commission reviews and opinion pieces for the "front-half" or magazine section of the journal and work with individual authors to improve their manuscripts for publication. A key part of my role is engaging with the scientific community by giving talks on the publishing process, doing site visits to meet scientists and discuss their work, and participating in initiatives that aim at improving the publication process and policies.

What was the focus of your PhD?

I did my PhD at the Research Centre for English and Applied Linguistics at the University of Cambridge back in the early 2000s. My doctoral work was in the field of psycholinguistics and used behavioral experiments to examine the psychological processes that underlie our understanding and resolution of ambiguity in connected speech. Specifically, my work examined how we identify the right antecedent for pronouns (e.g., s/he, him/her) in cases where more than one antecedent is available – how do we combine syntactic, semantic, and pragmatic information on the fly to select the right referent for an ambiguous pronoun?

As you were finishing your PhD, what were you thinking about your career plans?

Doing a PhD wasn't part of my initial career plan – but when the opportunity arose, I was very excited to have the chance to continue learning. As I was finishing my PhD, I had a very limited understanding of career possibilities outside academia – my eyes were firmly on an academic track because that's what I thought was the main option available to me. Happily, I was wrong, but it took 4 years of postdoctoral research before I actually found out!

Can you tell us a bit more about how you started to consider other options and became the editor of *Trends in Cognitive Science*?

I enjoyed my doctoral and postdoctoral work, but unlike many of my colleagues, I found it difficult to restrict my interests to a narrow field of research, which is both necessary and inevitable for early career researchers who want to stay in academia. I was very broadly interested in science and enjoyed learning, so I started looking into potential careers that would enable me to stay very close to science and continue learning but didn't involve specialization. At that time, I came upon an ad for the editorial opening at *Trends in Cognitive Sciences*. I had no idea that the option to pursue a career as a professional scientific editor even existed – I thought that all journal editors were senior academics who edited journals on the side while continuing to do research as their main job. However, when I read the job description, I thought I'd found my calling – the job profile had everything I was looking for and no previous editorial experience was required. I've never worked harder than I did on the application and test materials for that position, and luckily, I received a job offer!

Can you tell us a bit about what day-to-day life is like in your current position?

There is never a dull moment at this job, and I learn something new every day. My day typically starts by looking at the new submissions we have received over the past 24 hours, doing an initial triage, and assigning manuscripts to individual editors within my team (including to myself). I handle the editorial and peer review process for my own manuscripts and also weigh in on the decisions my team makes as one of our aims as a journal is to make consistent decisions regardless of who the handling editor is. We discuss manuscripts a lot as a team: we have dedicated triage sessions where the handling editor presents a manuscript and we reach an initial decision collectively on whether to peer review a manuscript or return it to the authors without review. Keeping on top of new developments is also a key part of the job. I scan on a daily basis what has been published in other key journals as well as new preprints posted on preprint servers. I check out press releases and what is being discussed on social media. I also attend talks, workshops, and conferences internally and externally to keep up to date with new research and science policy developments, as well as network. Talks, the news, and social media are frequently the source of commissioning ideas for our magazine section – I will often commission an opinion piece after seeing an interesting exchange on Twitter, for example. I also talk to scientists a lot – either to discuss their manuscripts, ongoing work in their laboratories, or interesting developments in their fields. My day may also include giving a publishing talk at a workshop, conference, or as part of a site visit; writing an editorial or a press release for one of our papers; developing a new journal policy; or working on a new publishing initiative with colleagues from other journals.

How do you think having a PhD has helped you succeed in your current position?

Having a PhD is a required qualification for professional editors of scientific journals – I would have been unable to pursue a career in publishing as a science editor without it.

If someone currently finishing their PhD was considering a similar position as you have now, how might they decide if it would be a good fit?

This job is for you if you like reading (a lot!) and making decisions, if you enjoy learning new fields and new skills and juggling several different tasks during the working day. Very broad interests in science are a must, as is a keen eye for the big picture and how different fields and disciplines relate to each other. Interactions with scientists are intellectually stimulating and rewarding, but editorial decision-making can also be challenging: publishing is a key vehicle for career advancement in academia, and dealing with unhappy authors whose work was rejected is a core part of the job. As an editor, you will need to be firm but fair, treating all scientists equally, regardless of whether they are a Nobel laureate or a PhD student submitting their first research manuscript. You need to be able to always maintain professionalism, as well as willingness to admit and correct errors. Being able to communicate effectively and network with scientific community leaders is also an essential skill.

If someone was interested in pursuing a similar career path, what would you suggest they do to better prepare themselves?

Read, read, read! Read as widely as you can, not just in your own and neighboring fields, but also in other disciplines and sciences. Sign up for eTOCs of general science journals and read sources of high-quality scientific journalism. Take every opportunity that arises to act as a peer reviewer – this is not only community service but also a valuable opportunity to see the editorial process in action. Build a strong network of contacts in your discipline and seek opportunities to collaborate on projects that extend beyond your core research topic.

What do you like most about your work?

I love the fact that I learn something new and exciting every day (I still can't believe that I'm getting paid to do something that I'd happily do as a hobby!). I also find the work incredibly intellectually stimulating – as a researcher, a substantial portion of my work involved running routine tasks or analyses that could feel repetitive or mind-numbing; there's nothing mind-numbing about editorial work, and my interactions with scientists at the cutting edge of their fields provide a constant source of intellectual stimulation. But perhaps most of all I value the opportunity to do work that has an impact on how science is carried out and communicated through journal policies, advocacy, and promoting research that is robust and has high social significance.

And what do you like least about your work?

The pace of editorial work can feel at times unrelenting – for instance, we don't have control over the volume of manuscripts that are submitted at any point in time. However, I generally thrive in fast-paced environments, so this is a minor quibble.

A lot of people like academia because they feel it gives them an opportunity to work on a topic that they deeply care about. Do you think this is also true in your current position?

Although editorial work doesn't allow you to work on a single research question you care deeply about, it does offer ample opportunity to work on multiple topics that are close to your heart. For instance, I believe strongly in the importance of meta-science and meta-research, so I encourage the submission of meta-scientific research to *Nature Human Behaviour*. I commission a lot of meta-science for the "magazine" section of the journal, and I develop policies for the journal that draw on the latest meta-science findings. I also care deeply about research that directly supports the sustainable development goals – again, this is research that I strongly encourage, prioritize for peer review and publication, and advocate for in my editorials.

Another reason many like academia is that they feel it provides them with more freedom than they think they would get in other positions. How much freedom do you feel you have to work on what you think is interesting?

As an editor, you have the opportunity to shape the journal you work on. For instance, when I was hired to launch *Nature Human Behaviour*, I shaped the journal according to my own vision of what its mission and identity should be. I wanted the journal to straddle the gap among the social, biological, health, and physical sciences and to be not just multidisciplinary, but the home for interdisciplinary and transdisciplinary research that struggles to find a home in disciplinary journals. I also felt and feel very strongly about rigorous, reproducible research – so I developed journal policies that support this vision, I encourage the submission of meta-science, and I have written numerous editorials that advocate for robustness and rigor in science and publishing. I wanted to redefine what constitutes a "significant scientific advance" in the context of a highly selective journal (not just novel findings but also replications, (dis)confirmations, evidence-based advances, and applied advances), and this is reflected in the papers we choose to send out to review and ultimately publish. For better or for worse, the journal's identity would have been very different, both in its scope and in its priorities and policies, under a different launch chief editor. Freedom to shape the journal you work on is true not only when you launch a journal yourself or are the chief editor of a journal. In well-managed journals, individual editors have the freedom to shape the portion of the journal within their remit, can take the journal in new directions by expanding its scope, and also contribute to shaping its policies. For instance, as an editor for *PLOS Biology*, I spearheaded the introduction of meta-science as a core discipline covered in the journal. I was also responsible for the magazine section and could put my own "stamp" on the range and types of content and opinions we featured in that section of the journal.

Based on your journey, what advice or suggestions do you want to pass on to someone who's currently finishing their PhD?

Doing a PhD is a substantial investment of time and energy. It's also the cause of lost income: you are still a student, likely counting the pennies, while most of your former schoolmates are earning and climbing the property ladder. Toward the end of your degree, you may be wondering if you did the right thing to invest in doing a PhD given that academic positions are limited and competition for them is fierce. However, having a PhD does offer a professional premium that is unavailable to first-degree and master's holders. More importantly, it opens many more exciting professional doors than you may think, and the fact that you are reading this book is a great step toward discovering some of them. Academia is only one potential path, and in fact, at least in the UK, more than half of PhD holders pursue a career outside science altogether right after their PhDs, with a further 17% securing non-university research positions in industry, government, or NGOs (see Figure 1.6 in "The Scientific Century: Securing our Future Prosperity," a policy report by the Royal Society, 2010). The range of possibilities for a rewarding, successful career is so much broader than before you started your PhD, which has equipped you with much more than subject matter knowledge. Take the time to think about your skill set and what you would most enjoy doing. Talk to PhD holders in different professions to gain insight into their roles and whether they would be a good fit for you. Attend career fairs and use any career resources your institution makes available to discover different career paths. If you are uncertain whether a particular job would be right for you, internships are an excellent option that will allow you to get first-hand experience of what a particular job would be like day to day. Ultimately, no matter what you decide to do, you will likely discover that your PhD was the most worthwhile professional investment you could have made – I certainly feel this way.

This was very informative. Thank you so much for sharing your experiences with us!

"Having a PhD Is Seen as a Strong Asset When Being Considered for Hire as an Intellectual Property Attorney"

Anastasia Greenberg

Anastasia Greenberg

Abstract In this interview, we hear about Anastasia Greenberg's path in pursuing a law degree after a PhD in systems neuroscience. Differences in intellectual property (IP) law in Canada and the USA are discussed, as well as the value of "soft skills," particularly communication. Practicing IP law can be surprisingly liberating and provide freedom to champion innovative projects that you believe in. A PhD at some IP firms is a requirement for being hired due to the technical and scientific nature of the work. Anastasia ends by discussing how both academic and alternative career paths are risky endeavors—a PhD is only the beginning.

Contents

A. Greenberg (✉)
Wilmer Cutler Pickering Hale and Dorr LLP, Boston, MA, USA
e-mail: anastasia.greenberg@mail.mcgill.ca

C. R. Madan (ed.), *Academia and the World Beyond*,
https://doi.org/10.1007/978-3-030-82606-2_13

Chris: Can you introduce yourself and tell me a bit about your current position?

Anastasia: My name is Anastasia Greenberg, and I am currently an associate attorney in the Intellectual Property Department in the Boston office of a "big law" firm. Our department consists of attorneys with "hard-science" backgrounds, many of us holding graduate-level degrees in fields such as chemistry, biology, electrical engineering, computer science, and the like. As a team, we work on intellectual property matters for our STEM-focused clients that include large companies such as big pharma and tech giants, startup companies, nonprofit organizations, and universities.

In my role, I engage in diverse sets of intellectual property tasks. I support clients with something we call "patent prosecution," which essentially involves writing new patents and making revisions as needed to obtain issued patents at the United States Patent and Trademark Office ("USPTO") and internationally as well. This work involves meeting with scientists and engineers to learn about the field of work and the inventions at hand, writing the patent (which is technically a patent application until it becomes a granted patent) including preparing figures, and corresponding with examiners at the USPTO when they issue rejections (e.g., conducting formal examiner interviews during which I present arguments as to why the invention claimed in the patent is useful and novel, while listening to the examiner's perspective and answering any questions the examiner may have to).

I also support clients with diligence work, which includes comparing existing patents owned by a competitor to a client's current or future technology to identify whether there is significant overlap and, if so, working on a strategy to support the client with challenging the validity of the competitor's patents, determining whether there are strong arguments that the client's technology does not infringe the patents, and/or supporting the client in finding ways to "design around" the patents to improve their technology and shift their focus away from their competition. Another aspect of my role is to perform landscape and competitive intelligence analyses, which involve searching and analyzing the patent and scientific literature within the client's technical field to identify "white spaces" where the client could focus new research and development efforts to differentiate themselves from their competitors through meaningful innovation.

Another major aspect of my work is patent litigation. In some cases, patent rights are disputed in court. Patent litigation is very complex from both a legal and scientific standpoint. It involves, for example, working with scientific/technical experts to understand the technology at issue inside and out, conducting expert depositions (i.e., obtaining sworn evidence), making sense of vast amounts of scientific/technical documents that will be presented as evidence, conducting legal research, preparing formal written arguments that are to be filed with a court or administrative body such as the US Patent Trial and Appeal Board, presenting on the litigation strategy to the client, and preparing for oral court hearings including holding formal "mock trials."

In terms of the specific scientific and technological areas that I work in, I currently have a diverse practice within the general life sciences space. About half of my work currently focuses on therapeutics. Therapeutics, biology in the biologics category such as nucleic acids, peptides, antibodies, large proteins, as well as drug delivery systems. The other half of my work is in the medical device space including molecular diagnostic systems and implantable heart repair devices. Engaging in such a diverse array of fields means that I am constantly learning on the job and no 2 days are the same. I am a lifelong learner, and my current role helps to feed my desire to constantly challenge myself intellectually and professionally.

What was the focus of your PhD?

I completed a PhD in neuroscience between 2011 and 2016 at the University of Alberta in Edmonton, Canada. My research was in the field of systems neuroscience and the neurobiology of learning and memory. More specifically, I investigated the role of brain activity during slow-wave sleep in "memory consolidation," which is the process by which memories become strengthened for long-term storage. My work was data analysis heavy involving signal processing analyses. My field was extremely interdisciplinary as it required a deep understanding of molecular/cellular physiology, as well as network-level physiology, the former falling within the biological field and the latter falling more within the fields of physics and electrical engineering. It is thanks to this interdisciplinary training that in my role today I am able to comfortably traverse both life sciences and technology/engineering-focused work (such as in the medical device space).

As you were finishing your PhD, what were you thinking about your career plans?

Near the end of my PhD, I first thought about pursuing a postdoctoral position. I initially started my PhD with the hopes of becoming a professor and a full-time researcher. However, I began having doubts about the academic career path for various reasons, and I became curious about careers outside of academia. I had a chat with a friend one night at a cafe who was also a PhD student at the same university, and he suggested that I consider a career in patent law. My initial reaction was that I would not be able to survive another 3 years of additional education. However, that friend ended up planting the seed that night, and eventually I decided that I would apply to law school. I began law school about a month and a half after I defended my dissertation.

Can you tell us a bit about what day-to-day life is like in your current position?

Although every day is different, my typical day starts with logging onto my computer to check my emails (of which there are usually way too many). I then look at my work calendar to see what meetings are scheduled for the day, and I also review my to-do tasks, which I usually keep in a living Word document that I update daily. My "quiet" work tasks for the day include perhaps doing some research on a specific scientific or legal issue; revising patent application drafts, memos, or formal written documents after receiving feedback from more senior team members; corresponding by email with team members such as other attorneys, paralegals, or my legal assistant or with clients; reading patents or scientific papers to prepare specific analyses; and preparing PowerPoint presentations for either internal or external (client) audiences depending on the issue at hand.

My independent work requiring concentration and quiet time is balanced by more social aspects of my job. I typically have at least 2 hours worth of internal meetings on any given day with other team members. At such meetings, we discuss updates on our work on the given client project, plans for building out our strategy, and next steps, and we will make collective decisions on key issues. In any given week, I will meet with many different individuals across different internal teams since we build each team for each client project based on relevant technical expertise and availability. On some days, perhaps 2 to 3 days out of the week, I will also have client meetings. Before the pandemic, clients would sometimes come to our office for meetings in which case we would book conference rooms or we would sometimes visit our clients at their place of work, for example, on-site at their biotechnology company. At client meetings, clients will typically prepare presentations to update us on their research and development results and future plans, or we will prepare a presentation to, for example, teach them about key patents that their competitors own and collectively make decisions on potential solutions.

In the middle of a typical workday, I head down a few floors of our office building to eat lunch at our firm's cafeteria, which has a fresh menu that changes daily. Some days we have a training or networking event during lunch.

What do you like most about your work?

There are many things that I like about my work: the independence that I am afforded, the high-level intellectual work that I am expected to constantly produce, the collaborative nature of my work, the ability to be mentored by team members who are at the top of their games, the ability to act as a mentor to others, the diversity of tasks that I engage in, and the career/job security that I have.

However, if I am asked to choose just one favorite part of my work, it is hands down the variety of scientific topics that I get to work on. I am constantly learning and working on new areas of science and technology. As an undergraduate student, I chose to study science because I loved science broadly, and I thrived on learning new topics and selecting a diverse course load for each academic semester. When I entered the latter years of my PhD program, I realized just how specialized I was becoming and how little I actually knew about other areas of science outside of my niche field. Now, I am able to once again indulge in a large buffet of scientific and technological offerings.

And what do you like least about your work?

Life is full of double-edged swords. My favorite part of my work—the diversity of scientific topics that I work on—inherently births the least favorite part of my work: my lack of direct contribution to research and innovation. Since I support clients working in diverse areas of research and development, my clients are the experts in their respective niche fields, and I am a generalist who does not contribute to their research in a hands-on way. Of course, the advice that I provide affects the course of my clients' research and development and business outcomes in a tangible way, but I do not participate in the undertaking of the research efforts themselves.

However, there are ways within the career path of an IP attorney to get closer to the science. For example, some IP attorneys choose to work in-house at a company or research institute, instead of at a law firm. Most IP attorneys who choose such a path usually transition from a law firm environment into a company after several years of experience working at a law firm. This is because law firms are seen as stepping stones to other career paths given their strong records in training junior attorneys. In-house IP attorneys work closely with the research and development team and can become an integral part of the innovation efforts. In this manner, an in-house IP attorney lives and breathes a niche technical field, much like an academic scientist would. I have also seen some IP attorneys pursue less traditional roles, such as taking up a position as an executive of a startup or more established company, or at a venture capital firm that invests in life sciences companies. Such roles can open up a variety of creative ways to use skills acquired through obtaining both a PhD and law degree.

How do you think having a PhD has helped you succeed in your current position?

Having a PhD is seen as a strong asset when being considered for hire as an IP attorney. At some intellectual property boutique firms (but definitely not all), a PhD is a minimum requirement for hire. Scientific and technical training is critical for my career path. I read technical documents and deal with technical issues on a daily basis. That is the heart of my work. If I do not understand my client's science/technology, I cannot write a patent application, I cannot provide advice on how to differentiate from a competitor's technology or why a competitor may be infringing my client's patents, and I cannot build a position on how my client's technology could be adapted or implemented in ways that are novel. All of the legal issues and arguments that I deal with are based on a scientific and technical understanding of the subject matter that I work on, and the subject matter is always a scientific/technical matter. That is not to say that there aren't brilliant IP attorneys who do not hold a PhD; however, since this interview is meant to highlight career paths for PhD graduates, I would emphasize that a PhD is not going to be wasted in an IP career path.

If someone currently finishing their PhD was considering a position similar to yours, how might they decide if it would be a good fit?

The role of an IP attorney is heavily focused on "soft skills" such as strong communication skills, in addition to the scientific "hard skills." Someone who is thinking about a career in IP should consider whether they are interested in completely leaving the bench behind to work in a leadership and support role for scientists and engineers. In my work, I have to do a lot of reading, writing, and speaking, whereas during my time in academia, those tasks took a back seat to lab work and data analysis. Personally, I knew that I preferred a role where I could learn, think, and speak about science without getting my hands dirty, if you will. On the other hand, some people really enjoy hands-on scientific work, and for those people, I would recommend a career as a scientist or engineer at a company as an alternative career to academia; these are the very scientists and engineers who I have the pleasure of learning from and working with as their company's IP attorney.

If someone was interested in pursuing a similar career path, what would you suggest they do to better prepare themselves?

The best first step is to reach out to currently practicing IP attorneys in your jurisdiction of choice (i.e., the country in which you plan to work) and ask them questions about their career path and everyday life on the job. Since you are thinking about entering a communication-focused profession, it is excellent practice to start networking as early as possible. You can try to reach out to current IP attorneys on social media platforms such as LinkedIn, attend relevant events at your current educational institution (e.g., check to see if there are open events at your local law school focused on IP), and even find IP attorney profiles on law firm websites and email them directly. In the professional world, it is customary to connect with new individuals. A stranger is a professional contact you haven't yet made.

The next thing to do would be to assess the status of the IP market in your own country. Since patents are not equally distributed across the world, some countries with a larger proportion of local patent application filings may naturally have more opportunities for IP attorneys. In addition, the IP market in a given country may be biased toward a specific technical field, depending on the dominant local industries. You will be able to get a sense of some of these issues through engaging with your new professional contacts.

Following an initial market pulse check, the next logical step would be to figure out the concrete milestones that you need to achieve before being qualified for hire as an IP professional. In the US, a common path for IP attorneys with PhD degrees is to start working as a patent agent at a law firm (either a general practice firm with an IP department or an IP boutique firm). This is a highly advantageous path as it allows one to try out IP law straight out of a PhD program without having to commit to the pursuit of a law degree. Many firms will allow someone to work as a patent agent for a year or two before asking them to apply to law school. Many firms will also pay the tuition for the patent agent who will typically work a reduced hours schedule at the firm while attending an evening law program. After completing the law degree, the patent agent would write the bar examination in their state and become an IP attorney (or patent attorney).

On the other hand, in Canada, it is almost unheard of to be hired as a patent agent and have an employer pay for a person to attend law school on a part-time/evening basis (at least I have never come across such an example). Instead, a person who finishes a PhD and wants to become an IP attorney would attend law school on a full-time basis and apply for employment at the same time as the rest of their law school cohort. The advantage of this path is that you don't need to juggle a high-pressure job and a high-pressure academic program simultaneously. It allows you to focus on your legal studies while getting to know employment options through formal employment recruitment that takes place at law schools. Given that I completed my legal education in Canada, this is the path that I pursued. I accepted the offer with my current firm over a year before graduation, which is unlike PhD programs where you likely won't have an idea of your next steps until much closer to graduation.

I completed my law degree in Canada but was to sit for the bar examinations in the US. This is another point for you to consider—depending on your jurisdiction, your employment options will differ as different countries have different rules on transferability of law degrees.

A lot of people like academia because they feel it gives them an opportunity to work on a topic that they deeply care about. Do you think this is also true in your current position?

Yes, definitely. This is one of the largest misconceptions about the academia/non-academia dichotomy. I too fell prey to the unchecked assumption that academia equals intellectual freedom while a career outside of academia equals "sell your soul for money." This could not be further from the truth. For example, even though I am a junior attorney, I get to choose my projects. I have a practice manager who is responsible for ensuring that my practice includes all of the types of work that I want to be building my practice around and as little as possible of that which I am not interested in pursuing. To some degree, there are limits on the options, which are set based on the work that is currently available and where there may be pressing needs. That being said, I have a lot of control over my career on a daily basis. Nobody breathes down my neck, and I define the shape and form of the final work product that I deliver to a client, with invaluable input from more senior team members.

I feel more liberated and have more agency now than I did as a graduate student. As a graduate student, I had one boss, my supervisor. Now, I no longer have one boss, and instead I am a part of many teams, working in a collaborative manner.

Even though as a professor you can work on a topic that you deeply care about, your freedom gives way at least to some degree to constraints such as whether you can get a certain project funded, whether you have the right graduate students to undertake a certain project, the current appetite of academic journals for particular types of research, and most importantly whether you will make tenure.

In addition to being able to work on a wide variety of topics that I care deeply about within my standard everyday work, I still have the freedom to contribute to academic scholarship and community outreach as well. Some IP attorneys juggle client work with writing scholarly articles, teaching university courses, organizing conferences, supporting junior attorneys through mentorship programs, leading professional societies and associations, and participating as members of a host of boards and committees.

Based on your journey, what advice or suggestions do you want to pass on to someone who's currently finishing their PhD?

My suggestion would be to ask yourself what factors you considered and what steps you have taken to evaluate your career options. In my personal experience, I found that most PhD students have only ever received career advice from their supervisor(s) and other academics. It makes little sense to receive advice solely from someone who has never built a career outside of the academic system, so it is important to cast a wider net when asking for advice. Also, consider that if you yourself have never had a career outside of academia, whether you're naturally gravitating toward what is familiar to you. You owe it to yourself to apply the same kind of systematic methodology to choosing your career path that you apply to your scientific research.

When it comes to scientific research, if you take big risks, you are more likely to fail. However, if you don't take big risks, you won't arrive at groundbreaking discoveries. The same logic applies to your career choice. You are just at the beginning of that journey when you've completed your PhD.

Thank you for telling us about your career path and perspectives, Anastasia!

The content of this interview reflects my own views and does not reflect the views of any organization with which I am currently, or have previously been, associated with.

"I Advocate on Behalf of Psychological Science Before the US Congress"

K. Andrew DeSoto

Abstract In our interview with Andy DeSoto, we find out about his path from PhD to policy. Andy works for one of the key psychology scientific organizations and frequently visits Capitol Hill to help the association advocate for federal funding or topics related to psychological science. Speaking and writing skills are essential to this job, and the credentials of a PhD are important for demonstrating expertise. This job allows Andy to not only stay in touch with his academic colleagues but also focus on making a difference on behalf of the field. He reminds us that career paths are often not direct, to seize the opportunities that arise, and to enjoy your time as a PhD student.

K. Andrew DeSoto

Contents

K. A. DeSoto (✉)
Association for Psychological Science, Washington, DC, USA
e-mail: ADeSoto@psychologicalscience.org

C. R. Madan (ed.), *Academia and the World Beyond*,
https://doi.org/10.1007/978-3-030-82606-2_14

Chris: Can you introduce yourself and tell me a bit about your current position?

Andy: Hello! I'm the Director of Government Relations at the Association for Psychological Science (APS). In this role, I carry out the association's mission by overseeing our connections with the government institutions that support psychological science. I advocate on behalf of psychological science before the US Congress and foster connections with the world's top science funding agencies and programs that support our research area. I also promote the application of psychological science to improve government programs and policies. Day-to-day this means I am often out of the office at meetings where behavioral science is being discussed or talking with research psychologists, program officials, or policymakers about developments in the field. I also spend time in the office researching policy proposals and strategizing and carrying out APS activities, such as issuing statements or sending formal letters. I also oversee APS's efforts to communicate policy developments and funding opportunities to our members. I am fortunate to be the lead psychological scientist on staff in our DC office, which means that I often consult and advise on our scientific initiatives.

What was the focus of your PhD?

My PhD is from Washington University in St. Louis's Department of Psychology. Roddy Roediger was my adviser. I started the program in 2009 and graduated in 2015. For most of my graduate career, I was focused on the relationship between confidence and accuracy in memory reports (see DeSoto & Roediger, 2014, for a representative publication). Near the end of my time in graduate school, I began to work with Roddy on issues of collective memory – how groups as a whole remember events of the past (e.g., Roediger & DeSoto, 2014).

As you were finishing your PhD, what were you thinking about your career plans?

As I was finishing my PhD, I was interested in a wide variety of potential career plans and engaged in a very broad search across many areas to try to find a potential career for which I'd be a good fit. The DC area is home for me, and my future spouse had just moved back to DC for work, so I was very motivated to establish a career in Washington specifically. I had "grown up" as a scientist with APS, so to speak, participating in their student programs and serving on their student leadership board, so I knew the organization very well. I'm grateful to APS Executive Director Emerita Sarah Brookhart for taking a chance on me and offering me a one-year fellowship to work at APS that eventually became my permanent role.[1]

As my PhD program was wrapping up, I explored other opportunities besides work with APS too. I applied for a range of US federal jobs on USAJobs.gov that I now recognize were totally out of my league or lane. I cold-emailed folks at tech startups and had several good conversations with folks doing applied psychological science at different places around the country. At one point, I even had a promising bite from a recruiter at Facebook, which eventually petered out as the APS momentum grew. The slow sense over time was that the stars were aligning for the next step to be toward APS.

Can you tell us a bit about what day-to-day life is like in your current position?

Day-to-day life is varied and a lot of fun. In the office, I spend most of my time in meetings or working on the computer, writing, or on email. But I also spend a lot of time attending local meetings and events. Given that Washington, DC is home to the US federal government, key science funding agencies are just a short Metro (commuter rail) ride away. Capitol Hill is a frequent destination. And DC is home to many think tanks too; for instance, across the street from the APS office are both the Brookings Institution and the American Enterprise Institute, to name a few.

One reason I like my area of work is because it's relatively fast paced and exciting. I might come into the office and learn about a new bill calling for US federally funded research in a particular area. I might strategize about how APS might craft a response to the bill sponsor encouraging them to integrate behavioral science, or psychological science specifically, into the work. I might then coordinate with my colleagues who oversee our news efforts or publications to ensure that the issue is adequately covered. I may then meet with individuals locally who are familiar with

[1] Be sure to read Brookhart's (2020) *Observer* column for an inspiring look back at APS's origins and trajectory.

the issue or attend a coalition member meeting where multiple folks in my line of work have convened to discuss.

It can be a stressful environment, sometimes, because at many of these meetings you may never know who you'll bump into, what you might be asked to speak about, etc. Eminent experts or sage policy wonks are encountered at every turn. When I began this line of work, most of my mental effort went into wondering whether I was over- or underdressed or which building entrance was the right one to take to get to a meeting. Fortunately, most of my time is now spent thinking about the actual subject matter of the meeting. (Nevertheless, I usually have a spare tie in my bag, just in case!)

For more of a sense of the kinds of things we work on at the APS office, be sure to check out some of my writing in our membership magazine, the *Observer* (e.g., DeSoto, 2021a, b).

How do you think having a PhD has helped you succeed in your current position?

A PhD has been extremely valuable. First, I find that a PhD education is a useful "advanced liberal arts education" that honed my thinking, writing, and problem-solving skills in a way that my actual liberal arts undergraduate education only started. But more importantly, understanding scientific research, and the research process, enables me to be the most effective advocate for the science of psychology. I have a better understanding of some of the challenges and joys members of my field experience – and if I don't, I'm a connection or two away from someone who does.

I'm happy to say that a PhD is respected too by policymakers and others – it helps when a congressional staffer, for example, knows you have a formal, doctorate-level academic background.

If someone currently finishing their PhD was considering a similar position as you have now, how might they decide if it would be a good fit?

I think it starts with writing and speaking skills, plus a sense of diplomacy. In this area, you have to have a good sense of what is the right thing to do and say in a wide variety of circumstances. It also takes a bit of a team mentality – a research professor, for instance, gets to oversee their own efforts and has a degree of independence in their work that is rare among other careers. (Or at least that's my view from the outside!) I have to keep APS's different stakeholders in my mind at all times – our members, first and foremost, but also our board of directors, which sets the vision for our organization; my supervisor at the office, the APS executive director; my colleagues; and more. Working at a membership society also has its unique challenges. I'd never want to publicly criticize an APS member, for example – those are the folks whom we exist to serve!

A lot of people like academia because they feel it gives them an opportunity to work on a topic that they deeply care about. Do you think this is also true in your current position?

Yes! I feel that I can enjoy the best of the academic and non-academic worlds in my current role. Working for a psychology organization that serves the field, I can keep in touch with many colleagues and friends from earlier in my academic career. I still keep up with some small research projects, I review papers and serve on an academic journal editorial board, and I attend conferences in my discipline. I feel like I've been able to maintain the positive momentum of my graduate career in the new role and that everything's been able to build off of itself.

Of course, not everyone in the scientific association world gets to work in their precise field, and I recognize my luck in this and in many other ways. At some point in my career, I may need to step into a workplace with more of a focus on health, or a different area of science, or one that deals with member needs that are different from those of academics. Fortunately, though, in the advocacy world, the abilities you gain in making a persuasive argument and anticipating counter-responses are widely applicable. But I do have a soft spot for psychological science.

Another reason many like academia is that they feel it provides them with more freedom than they think they would get in other positions. How much freedom do you feel you have to work on what you think is interesting?

There are trade-offs here, for sure. As mentioned earlier, as part of a team, I have obligations to many different groups of people. Reporting to APS's executive director, I have a boss, and in my workplace (as in most), you have to develop good relationships and understand what work is in your purview and what you must run up the flagpole. Just a day or two ago, I was contacted by a member of the media who wanted to talk about my scientific research. In a situation like this, I must do my due diligence in checking in to ensure I am cleared to speak to the media. Generally, there are no problems when it comes to this sort of thing, but I have to keep in the back of my mind whether my comments might (incorrectly) relay the views of my employer. Another lesson I've learned – slowly – is when to make a comment or weigh in and when to let other professionals do their work. A PhD does not make you an expert in marketing, or communications, or even scientific publishing!

This mindset also comes into play with social media. Before working at APS, I was a very avid social media early adopter and devotee. Recognizing the potential visibility of my position and how my responses might reflect on APS, I am now much more careful on social media and almost always prefer to retweet or reshare the thoughts of others rather than voice my own opinions on matters. As many times as I might like to wade into a particularly spicy methodological debate, I must think first!

To some people, these limitations may seem like a setback, but they haven't been too bad for me, for the most part. I appreciate having a clear chain of command in my workplace; it's great to learn from knowledgeable, talented supervisors and more senior colleagues; and it's a good rhythm to sometimes lead and sometimes follow. So it's been a fun adjustment for me, but it might not be for others.

What do you like most about your work?

I have the opportunity to attend many meetings and get a chance to hear amazing people speak. Often, when sitting in the audience or otherwise participating in an event on a particularly essential topic – for example, 2021's discussions about racism in society, or how psychological science can contribute to solving COVID-19 – it's exhilarating but makes you panic too. There's so much to do in the world, and psychological science has so much to say about so many societal issues. There are countless initiatives to launch, programs to spin up, workshops to hold, stories to write. I get to feel that excitement on a regular basis, that sense of seeing what science and scientists have to offer. The panic comes because the possibilities are truly endless. And in many cases – such as the racism example provided earlier – scientific and other communities have been tackling these issues for decades or more and have not made necessary progress. The easy, fun part is seeing all the opportunities for action. The hard part is rolling up the sleeves and seeing how you can contribute.

What do you like least about your work?

In policy, although the pace of work is quick, progress can be very slow. I used to think academic publishing was a lengthy process because it can take a few years for a standard research project to go from the idea stage to publication. Advocacy never really ends. Think of the priorities that US Congress has, for instance. Members of Congress are worrying about everything from military infrastructure to regulating social media networks. Given the many competing priorities, huge successes in this world can be small and far between, but the ongoing work is essential. It's possible that some years, only my colleagues and I are speaking to US policymakers specifically about the value of, say, basic psychological science. Someone needs to be sharing that message. I'm lucky to get to do that.

Another challenge I face is that, especially over the last 4 years, things in Washington have become even more partisan. People are angry – angry with their government, angry with their leaders. And you can understand why this is. But it can be hard to navigate the political system when topics are so raw and sensitive. I am working to understand this dynamic myself, to better understand when a quick, impassioned response to right a wrong is necessary and when a slower, more measured take is best. But there are many opportunities for missteps.

The last challenge – and it's an opportunity too – is that membership societies are in an interesting place now, with more discussions about science and communities forming online. All societies – not just ones in psychology – are grappling with how to serve their members and help their members recognize the value of membership and how to provide that value. What will scientific societies look like in 2030? Is there even a place for them? It's a tough problem to solve and one we're working on almost every day!

Based on your journey, what advice or suggestions do you want to pass on to someone who's currently finishing their PhD?

Privilege and good luck aside – which I recognize is asking a lot – I have three general pieces of advice. One is to think about what you want to do and different ways to get there. Someone who wants to do research may be able to do so in a variety of ways. Same for teaching, public speaking, outreach, and more. Many friends and colleagues of my academic "generation" seem to be making their dreams work for them, although the route they're taking isn't always a straight shot. A second piece of advice comes from a wise phrase I once heard from a faculty member at a Wash U department colloquium – "take what the terrain gives." That is, if opportunities present themselves, even if they are a little different from what you are expecting, it may be worth seizing them. You may find that the day-to-day joy of working in a career where you are appreciated and fairly compensated outweighs the future benefit of what you may see as a "dream job" now. If you have a unique opportunity or leverage, use it! Last ... do what you can to enjoy the last few months of your PhD. It is unlikely that you ever again will be surrounded by a group of people who have shared the same academic experience and training as you and who think scientifically in such a similar way. Although we psychological scientists know that hindsight is often rosy, I miss Roddy, my PhD adviser; good friends I made in the program; the collegial, challenging environment; and being over a decade younger just about every day!

Is there anything else you'd like to tell someone reading this interview?

As the reader knows, a PhD is a considerable investment, and it's not for everyone. But pursuing a PhD is one real way to make at least part of your life a vote for the way of knowing about the world that we call science. There's a sense of accomplishment that comes when you finish your degree – sort of a "Hey, I earned this, and now I have it for good." Although your road may be tough – and the journey is difficult for every individual in ways I can't personally understand – I hope you are proud of your choice to pursue the degree, or not, and I hope that this interview and the others in this compilation help in your thinking and decision-making.

Thank you for sharing your experiences, Andy!

Andy would like to acknowledge Kayla Burton and Rebecca Koenig for their helpful comments on the print version of this interview. As no surprise to the reader of this interview, this interview does not reflect the views of APS.

References

Brookhart, S. (2020). Back to the future: Why APS stands the test of time. *Observer, 33*(9), 5–9.

DeSoto, K. A. (2021a). Policy watch: Behavioral science momentum in 2021. *Observer, 34*(1), 21.

DeSoto, K. A. (2021b). Policy watch: Increasing attention toward behavioral science. *Observer, 34*(3), 27–28.

DeSoto, K. A., & Roediger, H. L. (2014). Positive and negative correlations between confidence and accuracy for the same events in recognition of categorized lists. *Psychological Science, 25*, 781–788. https://doi.org/10.1177/0956797613516149

Roediger, H. L., & DeSoto, K. A. (2014). Forgetting the presidents. *Science, 346*, 1106–1109. https://doi.org/10.1126/science.1259627

"Finding Your Passion and Staying Authentic"

Muireann Irish

Muireann Irish

Abstract In our interview with Muireann Irish, we discuss how she progressed from finishing her PhD in Ireland to her current role as Professor of Cognitive Neuroscience in Australia. Along the way, Muireann worked as a data analyst and saw how her PhD-related skills could be translated outside of academia, though she eventually returned to academia as a postdoctoral research fellow. Muireann also provides some insights into factors that contributed to deciding when to have a baby and how that aligned with career progression, but also raised her awareness of the leaky pipeline of academia. Since then, she has gained more experience in applying for grant funding, initially to create a small team, but later to consolidate a larger research group. The job role changes as you progress to becoming a group leader and increasingly work in a more managerial role. Academia should be viewed as a collaborative endeavour, including practices such as sharing previous grant applications and giving feedback to colleagues, as well as helping those who are junior scientists. As a trainee, it is important to try and have frank and open discussions with your supervisor, so possible future career options can be discussed.

Contents

M. Irish (✉)
School of Psychology, The University of Sydney, Camperdown, New South Wales, Australia

Brain and Mind Centre, The University of Sydney, Camperdown, New South Wales, Australia
e-mail: muireann.irish@sydney.edu.au

C. R. Madan (ed.), *Academia and the World Beyond*,
https://doi.org/10.1007/978-3-030-82606-2_15

Chris: Can you introduce yourself and tell me a bit about your current position?

Muireann: I am Professor of Cognitive Neuroscience in the School of Psychology, and based at the Brain and Mind Centre, which is a multidisciplinary research centre at the University of Sydney. In 2017, I established my own independent research group, the *Memory and Imagination in Neurological Disorders* group (*MIND*) comprising postdoctoral researchers, PhD students, honours students and research assistants from various backgrounds. We are interested in exploring how sophisticated and dynamic feats of human cognition are supported by the brain and how these processes are disrupted in neurodegenerative disorders.

What was the focus of your PhD?

I completed my PhD in Cognitive Neuropsychology at Trinity College Dublin in Ireland. My thesis explored how different facets of our autobiographical memories (i.e. our personal memories from the past) gradually change with the passage of time and how key features, such as the visual imagery and emotional experience of our memories, deteriorate in Alzheimer's disease. My PhD centred heavily around the development of a novel protocol to assess the phenomenology and the subjective recollective experience of autobiographical memory in aging and dementia, as well as to determine potential mechanisms driving these changes.

While I had not ever considered conducting a PhD, I had always been interested in memory. What began as a natural curiosity grew into a personal quest when my grandmother was diagnosed with Alzheimer's disease. Watching such an independent and spirited woman slowly forget our shared history was incredibly difficult, and I desperately wanted to understand what was happening. As I progressed in my psychology degree, I found myself gravitating towards the memory literature, immersing myself in the cognitive neuropsychology studies on rare amnesic cases. However, it was during the third year of my degree when the penny finally dropped. I vividly remember sitting in a Cognitive Neuropsychology lecture given by Professor Ian Robertson and realising that *this* was exactly what I wanted to study. I can remember the desk I was sitting at and the exact layout of the lecture theatre, almost as though I were back there today. It was a wonderful realisation that perhaps there was a way for me to integrate my long-held interest in memory with my new-found passion for the brain and to use this knowledge to better understand the cognitive landscape of dementia. In one of the many fortuitous strokes of luck, a paid internship was advertised to work on a study exploring the use of music to enhance autobiographical memory function in Alzheimer's disease. I was successful in applying for the position and turned this experience into my honours thesis, which then formed the backbone of my PhD.

As you were finishing your PhD, what were you thinking about your career plans?

To be honest, I was so caught up in trying to get my thesis submitted on time that I really didn't stop to think about my life post-academia. The deadline for my thesis was looming, and my funding only covered me for 3 years, so I scrambled to get everything finished and submitted on time. Following my submission, I was lucky enough to receive a small stipend to focus on writing up my various studies for publication, and it was during this time that I started to consider my next step.

How have your career plans changed as you've continued on to your current position?

My career has been anything but linear as I have had a number of deviations from the traditional academic path. Following the award of my psychology degree, I decided to take some time away from academia and spent a year working as an administrative assistant in an architectural firm in Dublin. Similarly, following the award of my PhD, I realised that I had been in formal education for most of my life and I wanted to take some time away to explore non-academic career options. I was fortunate enough to be employed as a data analyst at a large multinational marketing consultancy, and I spent almost a year working in this company. During this time, I also decided that I wanted to see some of the world and made plans to travel through Southeast Asia, China, Australia and New Zealand. Initially, my plan was to spend 1 year working in Australia before returning to Ireland to hopefully resume an academic career. This, however, coincided with the global financial crisis in 2008, which had left Ireland in a severe recession, resulting in a mass exodus of young graduates to Canada, the USA and Australia. It quickly became clear that I would need to stay longer in Australia, and I began to apply for postdoctoral positions in all of the major cities. During this time, I worked in two part-time research assistant positions that were not directly related to my primary research interests; however, I felt grateful to have relatively stable employment and an opportunity to potentially expand my research experience.

My first postdoctoral position came about as the result of a serendipitous email. I was scouting the major cities in Australia to see where the memory hubs were located and discovered that Professor John Hodges, formerly of Cambridge University, had relocated to Sydney and had established a large research group focusing on younger-onset dementia. I cautiously emailed him my CV, not expecting to hear anything back from him, and was pleasantly surprised when he responded and told me that they had just advertised a postdoctoral position to explore memory dysfunction in dementia! This was the opportunity I had been waiting for and would never have transpired if I had not taken the chance to email John. Thankfully, I was successful in my application for that position, resulting in another move from Melbourne to Sydney.

Can you tell us more about your post-PhD data analyst position?

While this was a complete departure from academic life, the data analyst position drew on many of the skills that I had developed during my PhD. Firstly, I had been unaware of how valuable the diverse set of skills that psychology graduates possess are and how such skills can be translated to the non-academic setting. When I interviewed for this position, I quickly realised that many of the basic inferential statistics that are the foundation of psychological research were readily transferable to this new setting. The fact that I could demonstrate a solid understanding of these principles and techniques, as well as being able to explain them in simple terms, was very important to the hiring panel. Interestingly, the company seemed particularly keen to recruit graduates with excellent interpersonal and communication skills, both written and verbal. These are skills that the vast majority of psychology graduates possess and demonstrated to me, at least, how much we have to offer beyond the immediate academic setting.

As a data analyst, I was working in a small team that managed accounts for large multinational companies. We were assigned to work on a specific company account for a set period of time, during which we would analyse their growth across brands, the impact of various advertising campaigns and promotional initiatives on brand activity, as well as how the launch of new products potentially cannibalised the overall brand growth. What was particularly interesting to me was how trends differed across different geographical markets and how we could use these insights to augment promotional activities. The position involved a lot of Excel and SPSS coding, along with regular team meetings and international teleconferences. There was a heavy emphasis on communication skills, and we underwent a number of training courses to improve our presentation style to various audiences.

Even though I had not envisaged working in a corporate setting, I really enjoyed the culture of the company. The management team assigned me to a mentor, we all played on a touch rugby team together in a tournament, and the company regularly organised lunches and dinners to celebrate group achievements. I did, however, struggle with the lack of autonomy in that the work proceeded along very prescribed lines with little opportunity to deviate from the set course. This approach is almost anathema to the innovation and discovery that is part of the scientific endeavour, and it took me some time to adjust to working in this manner. Moreover, as deadlines approached, late nights in the office were expected, and there was very little flexibility around this. I remember a particularly gruelling period where I had worked 14 hours per day straight for 2 weeks and was completely exhausted. The company provided meals for employees during such periods; however, this gesture soon lost its appeal. There is a distinct difference between choosing to work long hours on something you are passionate about and crunching numbers for a company report, and it was during this time that I started to feel my motivation for the position diminished.

Lessons that I learned from this time included the very valuable set of skills that psychology graduates can translate to the corporate/industry setting. We represent a rare combination of analytic and critical thinkers, with excellent written and verbal communication skills. As I progressed in the company, I became more involved with the recruitment process and assisted in screening applications and conducting interviews. It was very clear to me that non-academic recruiters are most interested in interpersonal and communication skills and that the fit of the person within the company culture is paramount. These insights have been valuable to me as I have built my own research team and have made me reflect quite considerably on how to build a diverse and collaborative team that has a strong and supportive culture.

How long was it between finishing your PhD and starting your first postdoctoral position?

From submitting my PhD to commencing my first postdoctoral position, there is a rather large gap of 2.5 years. This period of time includes my working in industry, then my relocating from Ireland to the Southern Hemisphere, as well as working in Melbourne in part-time research assistant positions for almost a year.

As you're a full professor now, what was your journey from your first postdoctoral position to where you are now?

My first postdoctoral position was working on a project funded by the Australian Research Council exploring prefrontal and temporal contributions to episodic memory, using neurodegenerative disorders as lesion models. This position required that I learn the basics of structural neuroimaging as well as familiarise myself with younger-onset dementia syndromes including frontotemporal dementia and semantic dementia. I had previously worked only with mild cognitive impairment and Alzheimer's disease patients, and so this was quite a steep learning curve but one that I readily embraced as I was so grateful to have made it back into academia. During the first 2 years of my postdoctoral role, I set about reading as widely as I could to understand how the episodic memory literature had evolved during my research hiatus, as well as developing experimental paradigms that I could use in the new patient populations I was working with. This was a really exciting time for me and a clear turning point in my career. I was given the freedom to pursue the research questions that interested me and to work with novel patient groups with rare disorders. There is nothing quite like the feeling of watching your hypothesis unfold right before your eyes, and I remember feeling incredibly lucky to have found such a supportive and thriving research environment.

As I entered the final year of my postdoctoral position, my supervisors encouraged me to apply for funding to enable me to hire my own research personnel and thereby establish some independence. I was extremely fortunate to be awarded a grant from Dementia Australia which enabled me to hire a part-time research assistant to support me with data collection and data management. This grant ended up being the crucial stepping stone to securing a large independent fellowship the following year. Having demonstrated that I could attract my own competitive funding and manage a small team, I was awarded an Australian Research Council Discovery Early Career Researcher Award (DECRA), which provided salary and research funds for 3 years.

As my DECRA was drawing to a close, I applied for and received a highly competitive Australian Research Council Future Fellowship, which gave me 4 years of continuous research-focused salary. This fellowship came at a pivotal point in my research career, where I had to make a critical decision regarding my future progression. The research team in which I had been based for the previous 6 years relocated to the University of Sydney, and I was given the opportunity to stay at my current institute or to relocate with the rest of the team. It was an incredibly difficult decision, but having just been awarded the Future Fellowship, I was strongly advised to negotiate for a permanent position at the new university and was promoted to associate professor in the process. In 2021, I finally made the leap to full Professor following a rigorous promotion application and interview process.

I feel like deciding when to have a baby and how that aligns with career progression is a topic that many struggle with. Can you tell us about some factors that should be considered?

This is one of the most challenging decisions that I have ever made. First, it is important to state that there is no perfect time to have a baby, and I think that can safely be said for any discipline or industry. There are so many factors that will influence this decision, and so I can only speak from my own personal experience in this regard. I knew that I wanted to be a parent but I had heard so many disheartening stories about the 'leaky pipeline' of academia and how women are inevitably pushed out of the system as it is incompatible with parenthood. Having worked so hard to get back into academia, I really didn't want my career to be derailed by something that should be a positive and exciting time in life. I entered into this decision acutely aware of the challenges that would lie ahead, while vowing that I would not become another casualty of the leaky pipeline.

A chief factor in arriving at this decision was my receiving an Australian Research Council DECRA fellowship, which provided me with 3 years of continuous research funding. Following conversations with senior women in science, I decided to incur the career disruption early in the fellowship and to hopefully regain my research momentum upon my return to work. This decision was made a lot easier by the fact that the university I was employed with at the time had a very generous maternity leave policy, enabling me to take 6 months of fully paid parental leave and to extend the end date of my fellowship without penalty. My research group was quite small at this time; I had one full-time research assistant, one honours student and a number of mentees under my supervision. In hindsight, this was both good and bad. The advantages were that I didn't have to balance the demands of maintaining a large lab group and ensuring my students were achieving their career goals during my maternity leave, but it also meant that my research productivity was impacted significantly. It was another 4 years before I was ready to have my second child, again timed strategically with another Australian Research Council – a Future Fellowship, which provides 4 years of salary for research. At my new university, I availed of an extremely generous package of 9 months of full-time paid leave; however, this time it was a lot harder to disengage from my research as my team had grown considerably and I felt a duty of care to ensure that my students and research staff were progressing well.

In reality, the term 'parental leave' is something of a misnomer as most academics cannot afford to take this time completely off and there are countless stories of new parents surviving on very little sleep as they struggle to write grants, submit publications and deal with the precarious nature of fixed-term contracts all whilst caring for a newborn. I don't endorse my way as the optimal way to decide on when to start one's family, but timing my pregnancies with the commencement of my research fellowships worked well for my particular circumstances and thankfully has enabled me to avoid becoming another statistic.

It sounds like you've managed to navigate the Australian funding landscape really well, particularly in having moved there after your PhD. Do you have any advice or insights related to relocating and navigating different funding systems?

This is a really important topic for junior researchers to think about prior to relocating to a new country. In hindsight, I was wildly naive about the Australian funding system and had moved there without doing any prior investigation regarding funding schemes or even the research culture and what is expected at different career levels. By the time I had commenced my first postdoctoral position, I was already playing 'catch-up' as the Australian academic system is heavily weighted towards publications. Indeed, many students submit their theses by publication, whereas in Ireland, the emphasis had typically been on producing the thesis with the assumption that papers resulting from the PhD scholarly work would be published during one's first postdoc position. I had five publications to my name when I commenced my postdoc and quickly realised that I needed to 'publish or perish' as the old adage goes; otherwise, I was not going to survive in this new system.

For the first 2 years of my postdoctoral position, I set about developing my own new research area – exploring the contribution of semantic memory to prospection and imagination by studying the rare clinical disorder of semantic dementia. In parallel with launching this new line of enquiry, I also analysed and wrote up publications based on existing data that had been collected by the team as well as publishing a theoretical piece in the esteemed *Nature Reviews Neurology*. My mentors strongly encouraged me to apply for small pockets of research funding, which, although time-consuming, eventually resulted in my being awarded a grant from Dementia Australia. I firmly believe that this grant, although small, was pivotal in my securing the prestigious DECRA fellowship from the Australian Research Council the following year. I worked closely with the Research Strategy Office at my university, attended all strategic sessions and workshops, researched the track record of previous awardees and managed to source copies of successful grants. This provided me with a clear roadmap of how to pitch my proposal and the areas I needed to strengthen to increase my chances. Even though I didn't have a large corpus of publications when I applied for the DECRA fellowship, I had a strong statement outlining my career interruptions, complemented by clear evidence of my upward trajectory and my growing independence in my research area. Now, when I review for these funding agencies, I am acutely aware of taking career disruptions and 'relative to opportunity' into consideration, as well as looking for evidence of the emerging independence of the applicant in their chosen specialty.

In essence, my experience has taught me that quality trumps quantity and that it is possible to make a strong case for research funding through a well-crafted narrative that demonstrates your growing research momentum. The most important lesson I learned, however, is that peer-to-peer support and strong mentoring are crucial. If a colleague of mine had not told me about the DECRA and encouraged me to apply, I doubt I would have ever considered this. Likewise, if other scientists had not been so generous in reading my drafts or sharing their own applications, my career pathway would look vastly different. My take-home message is that we must remember to be collegial and supportive, to inform others of opportunities, to share our previous applications and to offer to comment or review grant drafts. These small actions have such a powerful influence in determining the success of the next generation and can make all the difference in someone sinking or swimming in academia.

Do you think that someone who has just finished a PhD might not be aware of some aspects of being a professor and running a research lab?

Absolutely! One of the biggest challenges in running a lab is the responsibility of securing grant money to ensure that lab members have continuing positions and ongoing access to funding. I spend an inordinate amount of time writing grant applications, with the knowledge that if I am not successful, it could have grave implications for fixed-term members of my group. I find this part of the position quite stressful as other people's salaries and livelihoods are depending on my capacity to secure funding.

Another aspect of running a lab that many recent graduates may not realise is that as you rise through the ranks, you will be expected to commit a considerable amount of time engaged in service to the discipline and to the university. This can result in membership of committees from school to faculty level and the inevitable requirement to attend many meetings. For some, being able to contribute at the higher levels of the university system is hugely rewarding; however, others find the time spent away from their research frustrating. It is difficult to strike the appropriate balance of being a good departmental citizen and giving back to your university whilst ensuring your research group continues to thrive. Ironically, it seems the more successful you are as a researcher, and the higher you climb within the university system, you invariably become more removed from the research activities in which you excel!

If someone currently finishing their PhD was considering a position similar to yours, how might they decide if it would be a good fit?

This is an excellent question. I think there is probably quite a disparity between what emerging PhD graduates imagine a research career to be like and the reality of running a research group. For me, the critical factors to consider are communication and resilience. Making the leap from postdoctoral researcher to independent group leader is quite daunting as it means stepping out from the safety of your postdoctoral supervisor/advisor and forging your own path. As I mentioned, this means taking responsibility for funding your research team and ensuring you can maintain the longevity of the team over the years to come. With funding rates abysmally low in Australia at present, this is not an easy task, and I have found myself becoming increasingly adept at budgeting, planning, and cost cutting to ensure the group can continue its activities in leaner times.

I would also advise current PhD students to consider their own leadership and interpersonal skills before embarking on the path to group leader. Essentially, you will be assuming a managerial role, even though you may have never trained for a position of that type. Inevitably, there will be times where conflict arises and you or your team will question your decisions. In addition, it can be quite challenging to be the main point of contact for questions and troubleshooting – you may not always have the correct answer, and sometimes, you may not have any answers at all! Cultivating a strong collaborative culture can help immensely in this context, as well as being open and honest about your own strengths and weaknesses. I often find myself saying to students that I don't have the answer, but that we can find out together.

I would like to see a more open discussion of what a career in academia entails in terms of balancing administrative duties, grant writing, team leadership and research activities. It is really important that emerging graduates consider all of the 'non-glamorous' activities like grant writing and admin that are essential to keep a research team ticking along and to consider whether this is something they are willing to do as part of their research career. I truly believe that running a research team is one of the most rewarding and invigorating careers you can choose. Even through the more challenging times, the passion that I have for my research and the enjoyment that I derive from studying the brain easily make it all worthwhile.

If someone was interested in pursuing a similar career path, what would you suggest they do to better prepare themselves?

Even though my career path was somewhat circuitous, I gained valuable experience working across different settings and within diverse environments, which ultimately led me to realise that academia was the right fit for me. Although I didn't appreciate it at the time, I further gained a unique perspective on how to establish a strong culture in my team, as well as how to be creative and innovative in our research approach. As a result, I think it is very important for anyone considering a career in research to spend time working or volunteering in different lab settings and to gain exposure to different methodologies, techniques and approaches. If possible, conducting a research visit or placement overseas is a fantastic way to appreciate the many diverse approaches to science as well as to experience differences in research and lab cultures and funding systems. If this is not possible, then I would suggest establishing links with other research groups in your local area, even if their work seems tangential or on the periphery to what you are interested in. As I've come to learn, running a research group relies not only on your research prowess but on your ability to communicate, troubleshoot and manage, and you can gain valuable insights into this process by observing the research culture in all its manifestations across different groups. Above all else, finding your passion and staying authentic are essential.

What do you like most about your work?

I am enthralled by the complexities and mysteries of the human brain, and so being in a position where I am paid to think about and study the brain is my ideal job! I genuinely look forward to going to work every day, to learn from rare patients who generously donate their time to our research and to steer my team members towards hopefully finding a path that they are as passionate about.

No two days are ever the same as the work is incredibly varied and can range from conducting analysis, to writing theoretical pieces, reviewing papers or grants or assisting students with designing new study protocols or tests. Some of my most satisfying days, however, have been applying a new method in the clinic and watching my hypothesis unfold before my eyes. I have a number of very vivid memories of trying to keep my composure as I watched my predictions bear out in a testing session, then frantically rushing back to the office to score and plot the new data. I love watching my students get excited about their own studies and fostering this enthusiasm to ensure they have a rewarding research journey.

On a broader level, I sometimes like to step back and think about the global community of neuroscientists that I am part of and how we are collectively chipping away to try to understand the mysteries of the brain. It is a wonderfully humbling experience to attend conferences and meet some of the luminaries in the field, to establish new connections and to hear that our work is having an impact at the international level. Now, more than ever, the spirit of open science and data sharing means we can collaborate and join forces to tackle some of the big questions in neuroscience, as well as making friends across the globe. For me, there really isn't any better job out there!

And what do you like least about your work?

Aside from the obvious grievances about admin and grant writing, there are times when I do find my work affects me on a personal level. Working with patients who have progressive neurodegenerative disorders is hugely rewarding, but I am acutely aware that, at present, we do not have any viable cures or treatments for our patients. As we conduct longitudinal studies, we end up building up a rapport with patients and their families, and it can be quite distressing to see our patients decline in their cognitive and behavioural function as the disease progresses. One of the most difficult parts is when we receive notification of patient deaths, as many of us have our personal favourites who we have gotten to know over the years. During those times, I remind myself of why I entered this field – my passion to make a difference, no matter how small, in the lives of people living with dementia. I find my grandmother's memory is never far from my mind, and that gives me the motivation to persist during the difficult times.

Based on your journey, what advice or suggestions do you want to pass on to someone who's currently finishing their PhD?

Looking back on my career path, I realise how lucky I was to scramble back into the academic system and get to my current position without any clear planning or strategy. I think this is a case of being the exception rather than the rule, and I would strongly advise all final-year PhD students to not adopt this haphazard approach. The stress of writing up the PhD thesis can sometimes be so overwhelming that it overshadows everything else, but I think students also need to make a concerted effort to stand back to appraise their medium- to long-term goals, knowing that these are likely to change over time. Having frank and open conversations with your supervisor is also important – we cannot help you if we don't know what your hopes and aspirations are. I have had many interesting and important conversations with students where we simply talk about possible options, paths that might be interesting to pursue, and think about various academic and non-academic options that they might like to explore. This also enables us as supervisors to stay alert to various opportunities that we can recommend you for, as well as being able to discuss potential non-advertised postdoctoral positions with our international collaborators. And something I learned from my own circuitous trajectory is that the big decisions we make in life are actually rarely set in stone, and it is possible to regroup and find your way back to your rightful path.

Is there anything else you'd like to tell someone reading this interview?

Above all else, be authentic and be yourself. The archetype of the old male white professor has gradually been overthrown, and we are witnessing an inspiring movement towards diversity and inclusion within STEM. As a woman in science, I believe it's critical for the younger generation to see positive examples of successful female scientists and to demonstrate that it is possible to have a family and carve out a thriving academic career – one does not, and should not, preclude the other. As I gain more seniority in my role, I'm increasingly realising the importance of being kind. Science is often portrayed as a competitive 'dog-eat-dog' world, but some of my proudest achievements have come through collaborating with other scientists from around the globe. So as you ascend through the ranks, remember to look back every now and then, and try to make the path just that bit smoother for the junior scientists coming through.

You've shared a lot of great advice. Thank you so much for doing this interview!

"To Thrive at an Undergraduate Institution, One Must Love Teaching and Advising"

Jessica M. Karanian

Jessica M. Karanian

Abstract In her interview, Jessica Karanian tells us about her goal of staying in academia and finding a balance between teaching and research. After holding a visiting assistant professor position, she now works at an undergraduate-focused university. This role requires an interest in teaching and advising. Research in this setting is primarily done with undergraduate research assistants as volunteers or when enrolled in independent research project courses. Jessica provides some advice on how to prepare for this career path, particularly in engaging in teaching and advising when possible and in looking for professional development opportunities. She also reminds us that there are many jobs that involve stimulating, rewarding, and meaningful work—we just need to keep our minds open.

Contents

J. M. Karanian (✉)
Department of Psychological and Brain Sciences, Fairfield University, Fairfield, CT, USA
e-mail: jessica.karanian@fairfield.edu

C. R. Madan (ed.), *Academia and the World Beyond*,
https://doi.org/10.1007/978-3-030-82606-2_16

Christopher Madan: Can you introduce yourself and tell me a bit about your current position?

Jessica M. Karanian: In Fall 2019, I joined the faculty at Fairfield University as a tenure-track assistant professor of Cognitive Neuroscience in the Department of Psychological and Brain Sciences. The Department of Psychological and Brain Sciences offers a Bachelor of Science in Psychology and a newly launched Bachelor of Science in Behavioral Neuroscience. The department is composed of 11 full-time faculty members with active research programs that span many areas of psychology and serves over 300 majors and minors.

Fairfield University is a "more selective" private, primarily undergraduate institution with ~4200 undergraduate students. The typical teaching load in my department is three courses per semester, and the average class size is approximately 20 students with our largest classes capped at 30 students. Our mission centers around undergraduate education and the university has a very strong commitment to excellence in teaching.

Upon obtaining my Ph.D., I began postdoctoral training at Tufts University working on a collaborative project that spanned cognitive psychology and cognitive neuroscience. In 2017–2018, I was a visiting assistant professor of Cognitive Neuroscience at Wesleyan University. Then, in 2018–2019, I was a tenure-track assistant professor of Psychology at John Jay College of Criminal Justice with appointments in the doctoral programs in Psychology and Law and Behavioral and Cognitive Neuroscience at the City University of New York Graduate Center. I was lured back to my alma mater, Fairfield University, at the time when they were launching their new neuroscience major, because after my time in a research-heavy department, teaching one large 250-student lecture per semester, I realized how much my passion was focused on the undergraduate experience.

What was the focus of your PhD?

In 2017, I received my Ph.D. in Psychology with a concentration in Cognitive Neuroscience from the Department of Psychology at Boston College (Chestnut Hill, MA, USA). I worked with Dr. Scott Slotnick in the Memory, Attention, and Perception Lab. My graduate work focused on understanding the neural mechanisms that underpinned true and false memories for visual information using cognitive neuroscience techniques, including functional magnetic resonance imaging (fMRI), electroencephalography (EEG), and transcranial magnetic stimulation (TMS). Towards the end of my Ph.D., some of my research focus shifted to more applied topics, like eyewitness memory.

As you were finishing your PhD, what were you thinking about your career plans?

I seemed to always want to stay in an academic position. There truly has never been another job that has been of interest to me. While I briefly considered other options, I only did so as a "plan b" and truly hoped that I wouldn't have to resort to this plan. I began graduate school interested in both research and teaching. I was fortunate to have some meaningful teaching experiences as an undergraduate, serving as a teaching intern for statistics and a peer tutor for other psychology and neuroscience courses. I very much enjoyed these positions and found the role rewarding. So as a graduate student, I sought out teaching opportunities beyond serving as a teaching assistant, a role that primarily consisted of grading. This quickly led to my involvement in Boston College's Apprenticeship in College Teaching program. This program involved a series of seminars and workshops, as well as teaching observations and reflection exercises. This program also prepared and motivated me to teach three of my own courses during graduate school, two during the summer and one during the regular academic year. While I found it challenging to split my energy and focus between teaching and my research, I was confident that I was passionate about both and I was so glad that my Ph.D. advisor supported me as I explored my passion for teaching. However, the ultimate balance of how much time did I want to send teaching as opposed to focusing on my research was something that I was still unsure of while finishing my Ph.D.

For someone who has similar interests to you, what are some options you considered as part of your "plan b"?

I really had my sights set on a position in academia. I considered other positions beyond tenure-track assistant professor lines, such as full-time lecturer lines. There are quite a few similarities between full-time lecturer positions at R1 universities and tenure-track assistant professor positions at SLACs. Of course, there is less support for research as a lecturer (because it's not an expectation of the job) and also positions are typically renewed in 1–3-year contracts. I also thought that there might be other areas in the academic world that I could use my cognitive psychology/neuroscience training and apply that to teaching in a meaningful way. For instance, many universities have centers that support faculty teaching by running workshops and keeping faculty members up to speed on the "best practices" as shown by research. Clearly, I was really tied to the idea of being in a vibrant, student-focused environment as I always thought that would be the most rewarding to me. I did briefly consider whether I would enjoy high school-level teaching but decided it was not for me given the difficulty in specializing in psychology and/or neuroscience. Beyond academia, I'm not sure there were any jobs that truly excited me other than some unique roles at neuroscientific consulting companies, user experience roles, or within nonprofits geared toward criminal justice issues. Many people point to data science as an excellent alternative, and I would've considered it more seriously if my academic dreams did not come to fruition.

Earlier you mentioned that you had worked for a year as a "visiting assistant professor." Can you tell us a bit more about what that position is?

Visiting assistant professor (VAP) positions are often filled by candidates that have recently finished their Ph.D. and ultimately want to end up in a primarily undergraduate-oriented university. VAP positions differ between universities and sometimes even across different departments at the same university. In my role, I had a relatively light teaching load of 3-2 (three courses one semester and two courses the other semester). I know of many other VAP positions with heavier teaching loads (e.g., 4-4). My particular VAP position opened because two faculty members in cognitive neuroscience were on sabbatical that year. This is often the case for the creation of these positions, which unfortunately often means the positions are only for 1 year. Alternatively, other VAP positions are created in a time of transition—perhaps a faculty member retires or a need in a certain subject area has presented itself. These positions may be for more than 1 year and sometimes even turn into tenure-track lines. I asked about this on my interview for the job, and it was made very clear that this position would only be for 1 year due to the circumstances in the department. While I was still excited about the position, this was stressful because it meant that I needed to start applying for jobs for the following year during the first semester of the position. Still, because I wanted to end up in a tenure-track position at a primarily undergraduate university, I knew this extensive teaching experience would be worth it. During my short time in this position, I was also able to mentor an undergraduate student on an independent thesis project, finish up some of my own research projects from graduate school, and begin some new postdoctoral research. Also, one huge benefit that I did not plan for is that I had the option to count this time in the VAP position toward my tenure clock at my current institution given the heavy amount of teaching. Not many universities offer this benefit, but it is something to inquire about when on the tenure-track job market, especially at primarily undergraduate universities.

Can you tell us a bit about what day-to-day life is like in your current position?

In my current position, I teach three courses per semester. So during the academic year, much of my time is dedicated to teaching. I do also mentor students through research positions in my lab, for which they earn course credit. These students often help collect data from our participant pool. Some of these students choose to build on this work and complete an additional course in which they complete an independent research project. Advising these students during the semester on top of teaching three courses can be time-consuming, but the bonus is that it allows your research program to move forward and provides a truly rewarding mentorship experience.

I also advise approximately 40 students within the department. Fairfield University—like many SLACs—has a very hands-on approach where students must meet with a faculty academic advisor in order to register for the following semester. And I think it's worth noting that all undergraduate academic advising is done by full-time faculty members in the student's area of study, as opposed to staff members, as I know this is typical in some larger institutions. Advising sessions usually take place over a few weeks during the semester. These advising weeks can be hectic given the number of individual meetings, but I have found that it is a great way to get to know students and connect with them beyond the classroom.

When not teaching during the academic year, I am also attending department-, college-, and university-wide faculty meetings, as well as engaging in a variety of service roles. Such service positions include more department-specific activities like advising the Psychology Club or broader roles like serving on a committee that reviews university curriculum.

During the summer, we are "off contract"; thus, we do not have specific obligations. However, many faculty use this time to engage in their research. The university offers funding to support faculty research (both in the form of grants for supplies/equipment and supplemental stipends), as well as free summer housing to summer research assistants. Some faculty choose to teach additional summer courses as well. Furthermore, there are a variety of professional development offerings by our university's Center for Academic Excellence, as well as opportunities to engage with students during orientation weeks. These off-contract summer activities typically come with a stipend.

If someone currently finishing their PhD was considering a similar position as you have now, how might they decide if it would be a good fit?

To thrive at an undergraduate institution, one must love teaching and advising. It is perfectly fine to also love your research (I certainly do!). But if teaching and advising are not rewarding and enjoyable to you, then this job will simply feel burdensome. To figure out if you would like this, find opportunities to teach! Many Ph.D. students have teaching assistantships—so make the most of that experience. Ask the course instructor if you can give a guest lecture or host a review session before an exam. If your advisor is supportive and you can find the time, I also recommend teaching your own course. I was extremely fortunate to have an advisor who was fully supportive of my desire to explore my teaching interests, so I ended up teaching three undergraduate courses during my PhD. I am certain that this teaching experience made me a very strong candidate for a visiting assistant professor position and possibly viable even some tenure-track positions.

Another way to test out the undergraduate-focused academic path is to mentor undergraduates in the research lab. This is probably much more feasible than teaching independent courses during your Ph.D. You can start by inviting students to help with some of your own work and then graduate to advising them on their own undergraduate projects. As a graduate student, I had the pleasure of advising two undergraduate thesis projects, and I found these experiences to be incredibly rewarding and also a way to move some of my own research forward. I recommend asking these students for written feedback so that you can provide evidence of effective advising on future job applications.

Can you tell us a bit more about how you move your own research forward through undergraduate projects? For instance, how do you find what might be a manageable topic for a thesis project and how do you balance the research topic being in the students' own interests while still fitting with your own longer-term vision?

My university, like many, offers course credit to undergraduates to engage in research during the academic year, and provides some summer funding to support faculty-student research as well. Ideally, students will work in the same lab for multiple semesters, as it can be difficult to provide a rich, hands-on experience in just one semester (especially in labs that employ techniques that require intensive training). Students also have the option to develop and conduct an independent research project. These opportunities certainly help bring students into the lab. Once in the lab, students may express interest in big research questions that simply are beyond the scope of the professor's research interests and/or expertise. So, it takes some effort to guide the student to a specific, feasible research question that falls within the scope of the professor's research program. For me, this is the toughest part—because I often want to let students be creative and to think big—but doing so would come at a big time cost to my own research program.

Regarding the amount of training that is involved once a project topic is decided, there is a lot of variability in the amount of mentorship that will be required to get the project completed. Some students are incredibly independent and operate more at the level of a graduate student, while others require much more guidance.

Sometimes, this is a bit hard to predict, but typically students start in a volunteer or supervised research role and then only the most dedicated and intrinsically motivated students are advised to take on an independent research project.

Some more behaviorally oriented research studies are very feasible to complete with undergraduate research assistants. It is the more technical projects (e.g., EEG) that require many months of training. These projects are quite difficult to get off the ground in an undergraduate environment, but certainly possible if you find the right students and establish a pipeline of students to streamline the training process. While the quality of research is not impacted, the rate of research output is often slower at an undergraduate-focused university.

If someone was interested in pursuing a similar career path, what would you suggest they do to better prepare themselves?

There are a variety of things that I did to prepare myself for my career path. That said, I acknowledge that there are many paths that one could take to reach the same final destination. Regardless, the following certainly would not hurt:

- Engage in teaching and advising when possible, and begin as early as possible. As you engage in these activities, be sure to document them all. Think about your approach/philosophy and refine it as you gain experience and feedback.
- Look for opportunities for professional development. Many universities offer workshops and special lectures on the topic—some are even specifically geared towards graduate students. For instance, Boston College has an Apprenticeship in College Teaching program. The University of New Hampshire offers a variety of courses on college teaching for graduate students as well.
- Find a professor at your university that is particularly dedicated to undergraduate teaching and mentoring (ideally, someone who is not your Ph.D. advisor!). Many larger research universities now have full-time teaching lines (e.g., Lecturers, Professor of the Practice). These professors will be excellent resources for you. Ask if you can observe their classes. When comfortable, ask if you can give a guest lecture and have them observe you. Even better, perhaps do a couple over the years and have them document how you have developed as a teacher. Reflecting on teaching is a critical part of the process and something that faculty at primarily undergraduate institutions value.

What do you like most about your work?

There are many things I like about my work, but I think what stands out as the most unique is the freedom and flexibility. Aside from scheduled class meetings, I decide when I do my work. I am not "checking in" with anyone on a daily basis. I can bring my car to be serviced on a Tuesday and know that I can work later that evening or on the weekend to make up those hours. More specifically, I love that my research challenges me on a daily basis. I love designing new studies and learning about new findings or methods. I suppose this all stems back to my love of learning. As for teaching, I enjoy connecting with students and watching them learn over the course of the semester. Especially in my statistics and neuroscience courses, students often tell me that they are only taking my course because it's a requirement. Being able to get those students excited about anything from that point forward is incredibly rewarding. Related to the above point about loving to learn: teaching also provides endless opportunities for continued learning.

What do you like least about your work?

What I like least about my work is actually related to the above point about flexibility and freedom. While I think this is mostly a benefit of my job, it does come with some challenges. (1) First, once in a tenure-track position, the job can be a bit isolating and overwhelming. You previously worked with a team of individuals who were experts in your exact research area who could act as a sounding board for research ideas. However, in most primarily undergraduate institutions, you are the only expert in your specific area of the field, so your support system changes a bit. But, finding great colleagues and mentors in your department—even if they are outside of your specific research area—can really make up for that. You learn quickly that the problems you are facing might not be so unique to you after all. Still, I have found that maintaining collaborations beyond my institution and open communication with my former graduate school cohort and mentors, as well as postdoctoral advisors, has really helped fill in the gaps. (2) Also related to the flexibility, my work seems to have no boundaries. I cannot remember a week that I did not spend time working on evenings or weekends. As I progress in my career, I've come to realize that some boundaries are critical, and I think it is easy to become burned out without them. Examples of boundaries include setting aside a fixed time to unplug and do something for yourself (e.g., exercise) or limiting time on email to only certain hours of the day.

Based on your journey, what is some advice or suggestions you would want to pass on to someone who's currently finishing their PhD?

Keep your mind open! There are so many jobs out there that allow you to spend your days doing stimulating, rewarding, and meaningful work. These exist both within and outside of academia. And, remember that every job comes with its perks and downsides—even your dream job! Be open to learning about the wide variety of academic positions, which range from nearly 100% teaching to nearly 100%

research. Also, think about the way your job application will present to future employers.

Acquiring specific skills and experiences beyond your Ph.D. research can give you a big edge over other applicants. And remember that these other skills and experiences can be related to your PhD research! For instance, a few of my graduate school peers took time to become proficient programmers using online resources, like DataCamp. They used what they learned to enhance their own research while simultaneously gaining a highly marketable skill. Also, many of my peers taught upper-level undergraduate courses in their Ph.D. subject area. For example, I taught a course on Human Memory (my Ph.D. was on the neural underpinnings of false memories). Teaching this course provided an opportunity to further explore and master my research area while also providing excellent preparation and exposure to college teaching.

Thank you so much for sharing your experiences and perspectives with us!

"I Really Enjoy the Hectic Multitude of Jobs I Get to Complete in Any One Day"

Gavin Buckingham

Gavin Buckingham

Abstract Gavin Buckingham tells us about his research and current academic position in his interview and shares some lessons he learned along the way. He suggests that getting experience with the academic roles beyond research, such as teaching and supervising, can be helpful in determining if academia will be a good fit. Gavin also moved after his first faculty position, and we discuss factors that contributed to this decision. PhD students do not need to define their career based on their doctoral research but rather consider that their environment and colleagues will change as they move along, and this can be used as an opportunity to change the research as well. Becoming involved in university-level committees early on can also be a useful way to gain a better understanding of university decision-making.

Contents

G. Buckingham (✉)
Department of Sport and Health Sciences, University of Exeter, Exeter, UK
e-mail: G.Buckingham@exeter.ac.uk

C. R. Madan (ed.), *Academia and the World Beyond*,
https://doi.org/10.1007/978-3-030-82606-2_17

Chris: Can you introduce yourself and tell me a bit about your current position?

Gavin: I'm a cognitive psychologist by training with an interest in how humans combine their sensory information with prior knowledge to experience and interact with the world around them. I'm currently a senior lecturer in the Department of Sport and Health Sciences at the University of Exeter, where I'm part of the Human Movement Science research group and I head up the Object Interaction Lab.

What was the focus of your PhD?

I conducted my PhD at the University of Aberdeen in the Psychology department where I did my undergraduate studies. My PhD supervisor (Dr. David Carey, now at Bangor) was also my BSc dissertation supervisor and got me really interested in the question of what it means to be right handed and why left handers exist! My PhD focused on the question of how people control their attention, with its single point of focus, during tasks which require you to use both of your hands at the same time (which, when you think about it, is almost everything you do with your hands!). There is an old idea that you are biased toward the rightward hand's task (or the right side of space) when performing these bimanual tasks, and this asymmetry might be what underpins the eventual journey to later-life right handedness. I used behavioral motion tracking studies to try and infer where attention was during tasks where you had to point toward visual targets with each hand simultaneously, finding some evidence that attention does shift toward the right during bimanual tasks for right handers. Left handers show no such asymmetry, which is pretty much what you'd expect from the theory (if you are really interested, let me know!)

As you were finishing your PhD, what were you thinking about your career plans?

I was pretty sure I was going to do a postdoctoral position somewhere. There was nothing else on my horizon except for an academic career. I applied for a few positions without any luck and then stumbled across a Commonwealth scholarship which would provide a year's worth of funds to work in a Commonwealth country. Handily, my PhD supervisor's PhD supervisor was a professor at the University of Western Ontario in Canada. I wrote the application but just missed the cut. Luckily, someone else above me dropped out and I got their spot, so off to Canada I went!

Can you tell us a bit about what day-to-day life is like in your current position?

Incredibly varied! The one constant is meetings – meetings with groups of colleagues as part of a committee (I am a member of about five committees, each of which meets around once per term), one-to-one meetings with PhD students (which I try to do weekly), meetings with undergraduate dissertation students, meetings with students who want help with their coursework, ad-hoc meetings with colleagues and/or the boss. So even this "constant" changes all the time. It's quite satisfying though to be able to jump from one task to another and feel like you are somewhat important to a range of different things. In between meetings I'm preparing for the next meeting, getting ready for teaching, dealing with email inquiries (usually from students), working on writing papers/grants, or fiddling with some technical job in the lab (the ultimate guilt-free procrastination). While this might not seem that satisfying, I really enjoy the "keeping all the balls in the air" aspect to the job and staying (just about) on top of everything that I need to do. Balls, of course, get dropped, but it is rarely a major problem for anything other than your self-esteem.

If someone currently finishing their PhD was considering a similar position as you have now, how might they decide if it would be a good fit?

I'd ask if they've had a chance to experience some of the other aspects of the job – namely, teaching, marking, supervising, and organizing. Typically in a PhD, there's not much exposure to the wider world of academia outside of your thesis – maybe the supervision of the occasional undergraduate project student, but that does little to prepare you for the day-to-day life of an academic which does, at first, feel incredibly chaotic. But if you are able to manage to ask for help when needed and are able to notice when you are sinking a bit, those are important skills to have for the job.

What are some aspects of the job that you think a PhD student would be particularly unprepared for?

First, how long it takes to prepare teaching. If you are preparing something from scratch, even in a topic you know back to front, it's a real challenge to put it into a format which (1) fits the time you have to fill and (2) is something more interesting than 44 hours of you lecturing in front of a PowerPoint of words. Adapting someone else's teaching is almost worst – second guessing what they'd say, disagreeing with their structure, and trying to figure out the narrative that goes with their visual. I'd always recommend starting from scratch. The second major issue is that you are, functionally, no longer a researcher in the same way you were before. You don't have a project. You oversee a load of projects. Some of these you won't be all that invested in, but they still require your full attention. So you learn the requisite knowledge in a far shallower way to have enough time to do all the other things you have to do. Finally, as alluded to above, the shifting of mental gears from one project to another project to an administrative task to a stats question.... This is really quite a change from (at least my) PhD experience, where I was able to just get my head down and focus on something all day long. Days like that are fleetingly rare for me since starting as a lecturer.

I know you've written some blog articles giving advice for new lecturers/assistant professors, can you tell us a bit about those?

The blog started off when I moved to my second lectureship and made me realize that it was a lot easier the second time around to start again without so many of the worries that plagued me the first time around. So I first started documenting to myself what I thought didn't go well the first time around in terms of swallowing up time and mental energy (mostly boring stuff, like documenting hardware setup). Then I noticed that I actually had lots of thoughts about this. Lots. And I actually really enjoyed writing this sort of content – no need to worry about drafts or proofreading or a narrative or anything like that – just the ideas in my head falling out of my fingers. So I started off just writing about what you have to do to set up a lab and then going through my own experiences of the other really difficult parts of starting out as a lecturer (coping with the supervisory responsibilities of a large number of undergraduate dissertation students, supervising PhD students, applying for funding, etc.). But because I love giving advice to people, it inevitably transitioned into a more general-purpose blog aimed at early-career academics of all stripes, on topics which I felt I had something to say. Some of these were very reactive to a situation I had just been through which wasn't really part of the job description (e.g., the huge amount of pastoral care duties you find yourself undertaking) or to questions which a colleague had asked me ("how do you prepare to be the internal examiner for a PhD thesis?"). It's been a great experience though, that got me invited to give a presentation at a workshop for early-career researchers and to write an article for *Times Higher Education* – both things which I mention on promotion, funding, and fellowship applications fairly regularly.

It would be great to hear some of this advice! It would be great to get a summary of a few of these, along with details on where to find the full articles. Can you give us an overview of workload and the responsibilities that go with being in a faculty position?

The main shift from PhD student or postdoctoral research to faculty is the responsibility and decision-making aspects. When you're only working on your own project, you have to make decisions about your own research. As faculty, you're making or helping others make decisions about their own research projects, sometimes on a small scale (design choices for individual research projects for undergraduate students) or at a large scale (what direction should this PhD take after this particular finding). You'll probably be leading a course or two, which you'll have some degree of autonomy over the content, structure, and assessment. You're also now a cog in the machine of the structures of your department and your wider institution. You probably have an administrative role, where you make decisions relating a diverse range of things from ethical approval of research projects to ways to the recruitment of students from disadvantaged backgrounds. There's not much in the way of a safety net either – in most cases, no one will be screening your decisions, and it rapidly becomes unviable to ask for advice in every instance. You're on your own, with no formal training. But you actually have probably got here by being reasonably decisive anyway. And you learn by doing. In my experience, the environment has been friendly and forgiving enough that there are consequences when things don't go as expected.

The article about this, and other topics, can be found on my blog: https://gavinbuckingham.wordpress.com.

What is your approach for supervising a large number of undergraduate students?

With difficulty! As a postdoc, I'd overseen the project of one extremely good student a year. As faculty, that number went up to around eight, with varying degrees of enthusiasm for conducting a research project (in the UK, an honors research project is a mandatory part of most science degrees). Initially, I'd have a whiteboard in my office with all my students' names and a grid of all their key milestones (project proposal, ethics application, data collection, etc.) so that I could tick them off and keep track of their progress. Now I'm using the same project management software I use with my PhD students (Basecamp) to keep track of their projects – this means I can send messages to them as a group and also have individual chats with them. Email contact is just a bit too unwieldy for the mixture of trivial and complicated queries which make a dissertation project. Also, while I like my students generally to be working on individual projects, I like to bring them together at least twice – first a few weeks into term to give each other an overview of their projects and then a few weeks before submission of their first drafts to present their findings to one another. Some students don't engage at all with this aspect (which is fine), but some really seem to benefit from this introduction to lab culture. Finally, try to be practical about your own workload. Initially, my aim was for all students to undertake a significant research project which was in my wheelhouse. These days, I let students have far more choice in what it is they want to do which frees up a lot of lab congestion – particularly critical now that I have a few PhD students. Something this year I'm trying is also to have a student supervised by each of my later-year PhD students on an "arm" of their current work – I really enjoyed this experience as a PhD student, so I'm hoping it will work out nicely from this more meta-perspective.

What is your advice for supervising your first PhD student?

Tough to give one-size-fits-all advice here because every student is so different – you want to try to be flexible enough to adjust to their needs (which themselves might vary over time) while teaching them the technical skills, methods of project management, and writing styles which you think will make them better scientists. As I've gone on in my career, I have become generally more hands-off with regard to the oversight of the project itself (often rarely more than a sounding board for the decisions they have made themselves) and more hands-on with making sure my students are doing ok in the bigger scheme of things. I now tend to have weekly individual short meetings (longer if significant project content needs discussion) and a weekly group lab meeting, as well as regular chat-style interactions throughout the week. And really a lot of this discussion is social rather than work. This pastoral care element I think is really critical because your goal as a supervisor is to get them through the process more prepared for a life as a scientist than they were before, rather than to mold them into new and improved versions of yourself. I know plenty of people who came out of their doctoral training with a ton of publications, but no desire to continue on. All that said, don't neglect their training in the early days when they have capacity. Things I'm particularly keen on making sure trainees pick up some skills in are (in no particular order) programming, good data management, good data visualization, clear and easy-to-understand writing, and open science. Note that all these skills are very highly regarded outside of academia, which is becoming increasingly key. Many (statistically, the majority) of your trainees will end up in careers outside of academia, and this should not be viewed as a failure on either your part or their part. Most important advice I can give though in this regard is not to heap expectations based on your own experience and trajectory (which, if you have got to the stage of supervising PhD students, is pretty exceptional). You don't need to push them to succeed. You need to support them in whatever their definition of success, inside and outside of the lab, happens to be. They are not a resource to be utilized, but individuals you are supposed to be training.

What do you like most about your work?

I really enjoy the hectic multitude of jobs I get to complete in any one day and the sense of pride I get from keeping all the balls (just about) in the air. I also really enjoy the mentorship of students – particularly my PhD students.

And what do you like least about your work?

I think that the reward structures are rather poor (although possibly not poor compared to the rest of society, but that's a conversation for another day). But a lot of the promotion in UK academia hinges upon income generation rather than the generation of knowledge. I'm pragmatic enough to understand why the university is interested in income generation but feel that absolute amount of money won, in an ecosystem where there's often a 10% chance of success, does not motivate academics to become well rounded (or happy!). I'd guess that promotion based on overall job performance (however we characterize that), including funding applied for rather than funds generated, would have a multitude of good effects on the academic ecosystem.

You currently work at the University of Exeter, but this isn't where you first were hired as faculty. Can you tell us a bit about why you looked for another faculty position?

My first position was in a historically small department which was being grown rapidly. I was hired as part of a push to hire research-intensive academics to boost the department's research ratings. This had many advantages – I didn't teach as much as some of my colleagues, and I had a pretty hefty startup package (at least for a UK academic, where we don't fare particularly well). But as we went on, I became frustrated with how the university was being managed at multiple levels and was left with the growing sense that the good times were over with regard to my protected position, particularly as they'd been making it pretty clear to me that I was underperforming in their eyes (i.e., not bringing in enough grant income). A promotion attempt got knocked back without any concrete description of what I'd need to do to get promoted, and at that point, I decided it was time to start looking around. At this point, my partner was also keen to move to a (slightly) warmer part of the world, and our child was young enough that we felt that a move wouldn't be disruptive at all. So the barriers seemed few, and it was exciting to try to imagine other potential future lives. I had a few interviews without success (in which I felt I learned a lot quite quickly) and then one which was successful before also being offered my current post. When offered the job, it was a scary proposition – not so much the move itself (which would be considered an upgrade in terms of university rank, for what that's worth), but moving out of my home discipline (psychology) into a new one (sport and health sciences). It's been an unambiguously good move for me though, and I'm incredibly happy with my life and work.

Thanks for telling us more about that process. I think that aspect – considering moving to another university – is particularly mystifying for those still figuring out what it's like to be in a faculty position. Can you elaborate on some of what you learned from those unsuccessful interviews?

I think that the main mistake I made during my unsuccessful interview was that I simply answered the questions that were asked. What I should have done was to use the questions asked in the interviews as a mechanism to demonstrate my suitability for the post, fit with the department, and knowledge of the university sector. By the time I got this right, I felt like I was able to give reasonably elaborate and knowledgeable-seeming answers to the questions being asked. I also became a lot more open – one of the reasons I wanted to move was that I was genuinely excited about the opportunities that would be presented at these other places. And letting the interview panel see that enthusiasm probably didn't hurt. This, of course, gets easier with experience – it's a lot easier to move "laterally" (from one lectureship to another) than it is to get your first position, so as things went on, I felt more confident, even with the unsuccessful ones.

Based on your journey, what advice or suggestions do you want to pass on to someone who's currently finishing their PhD?

Don't feel that your PhD (the experience, your workload, your stresses, or even the research you have been doing) needs to define your career. My time doing a postdoctoral fellowship became far more relevant to the research I do these days, as well as broadening my horizons and giving me a wider sense of what an academic career could be like. Many people come out of their PhD thinking that they certainly don't want to spend any more time doing that. But rest assured, you won't. The environment will change. Your relationship with your supervisor will change. Your responsibilities will change. Even your research will change (if you want it to). One of the great things about academia is that, at all levels, there are opportunities to shake it up if you are getting bored and disheartened with what you spend your day doing.

Is there anything else you'd like to tell someone reading this interview?

I think that it can be very easy to feel like an unappreciated cog in a large machine, whatever career stage you might be in. For me, this was particularly obvious when I was starting out as a lecturer – a huge number of tasks are pointed your way without a great sense of what their purpose might be. But getting more involved in higher-level institutional opportunities (joining the Senate, feeding back into focus groups, etc.) can be a really valuable way to get an insight into the drivers of the university – what makes them tick and why they make the decisions they do. Because in the UK universities are not profit-making enterprises, finding out the narrative behind some of the more cutthroat-seeming decisions is surprisingly relieving. And as an early-career researcher, you'll find that your views are more valued than you'd expect at college or university-level committees – the people at the top of a university are aware that there's rarely a "voice" for the early-career individuals at these committees but typically unable to find people – so put yourself forward and/or seek out these opportunities.

Thank you for all of the advice and perspectives. They are much appreciated!

"Find a Mentor, Someone Who Loves to Teach and Who Is Good at It"

Kelly J. Arbeau

Kelly J. Arbeau

Abstract In the interview with Kelly Arbeau, we discuss how she has adapted to adversities—including PhD supervisor leaving and university closing. Despite these challenges, Kelly found fulfilling job roles and was able to secure a tenured academic position. In this process, Kelly also describes how she revised her strategy for how she fits within her department and university. Consider how much you enjoy teaching and what other considerations may relate to where you would want a position, such as location, but also faith and other potential values of the university. Deciding your own priorities and expectations related to collaboration, teaching, and research can help you make sure you stay on track with all of the roles that need to be balanced. There are many great resources out there, from workshops to books; use what you can to prepare yourself for the job market.

Contents

K. J. Arbeau (✉)
Department of Psychology, Trinity Western University, Langley, BC, Canada
e-mail: kelly.arbeau@twu.ca

C. R. Madan (ed.), *Academia and the World Beyond*,
https://doi.org/10.1007/978-3-030-82606-2_18

Chris: Can you introduce yourself and tell me a bit about your current position?

Kelly: I am a tenured assistant professor at the Department of Psychology at Trinity Western University. I recently ticked the last box on all the things needed to apply for promotion to associate professor, 1 year ahead of my personal goal. Faculty at TWU are required to apply for tenure at the 5-year mark, but we have the flexibility to largely set our own goals with respect to promotion.

My interests include integrating quantitative (especially multilevel modeling) and qualitative methods in interdisciplinary health research, as well as the phenomenology of involuntary separation. My third and more surprising (to me) area of research interest is motivation for playing video games. Several years ago, a former student approached me wanting to do something about gaming and game players, and this "fun side project" was born. I think there can be real value in pursuing side interests in research. Over time, the gaming project has grown: two good friends who are graduate students at other universities have joined the team as collaborators, two undergraduate students have received funding to support their own work on the project, and recently two faculty colleagues and I received some funding of our own. (As an aside, am I a gamer? I am now!) I did pick up a replacement "fun side project" too, research relating to creativity and the arts. We staged an art exhibition last year that communicated the results of a small research study on creativity and religious faith. It was the experience of a lifetime. I can't draw worth beans, so I composed verbatim poetry to illustrate each of the study themes. The team lead, an art and design professor, recently pivoted his focus and invited me to join himself and a biologist on a project aimed at developing a set of creative practices for scientific research and knowledge mobilization. A few days ago, we learned that we have been awarded a federal grant to support this project. Our "test case" focuses on new management methods for invasive knotweed in British Columbia. Research is 30% of my contract.

I also teach six undergraduate courses/year (60% of my contract), including the junior and senior research methods courses, social psychology, cognition, and psychology of gender, plus occasional offerings of other courses, such as health psychology and intro psych. Because we are a small department, I am sometimes asked to teach outside of my area of expertise. When that happens (e.g., psychology of gender), I commit to reading wide and deep, and to consulting with experts, to make sure that the course is delivered to a high standard. I still strongly prefer to have at least some graduate-level training in any course that I teach, however.

In support of the service component of my contract (10%), I act as Tri-Council (Canadian federal funding agencies) ethics expert on the human research ethics committee and sit on the Undergraduate Academic Council, which reviews program and course changes. I also supervise undergraduate honors research and co-chair the psychology research participation program, which I was asked to co-develop and introduce in my first year at TWU.

What was the focus of your PhD?

I defended for a PhD in health psychology from the Department of Psychology at the University of Alberta in 2007. Those familiar with the University of Alberta will note that the department does not offer a PhD in health psychology. I was initially slated to work with a new hire who held strong health research interests. She moved her lab to another university about 6 weeks before my move to Alberta. I'd already confirmed my spot, secured a room in the most beautiful residence on campus, Pembina Hall, purchased a plane ticket, and (the most important factor) the university's offer of full funding was still available, so off I went, sans supervisor with health interests. The department and my replacement supervisors (one in psychology, one in public health) generously facilitated my efforts to patch together a *very custom, very multidisciplinary* PhD program.

As you were finishing your PhD, what were you thinking about your career plans?

I was career planning long before I realized that the PhD program was itself part of my career. Within a year or so of entering the program, I started working toward a certificate in undergraduate teaching. It was actually my mom who recommended it; I had little interest at the time in doing too much beyond the already quite rigorous program requirements. I taught my first course shortly after passing my candidacy exam in 2005, which was the department's requirement for course instructors. It was a spring (May–June) section of introductory psychology, and at age 25, I am pretty certain that I was the youngest person in the room. I was so intimidated by the whole situation that I wore a suit to every class. My hands shook a lot, my voice quavered, but the students were so kind and so patient, and they feigned surprise at the end of the semester when I said it was my first time teaching. I even made the teaching honor roll!

As I was entering the latter stage of my PhD program, I was thinking about settling into a position at a small, undergraduate-focused university and having a cozy little straightforward career path from assistant professor on up. I was deeply concerned about the research productivity part of the contract, but more or less confident that I could work something out. I started doing phone interviews and campus visits at this point.

And as I was finishing my PhD, I was in the latter stages of interviewing for my first faculty position, which I was offered shortly after my December 2007 defense.

Little did I know that "enrollment is booming" was one of those stories that people tell themselves and their job candidates when they want to pretend they're not actually on a sinking ship: I acquired a condo and a mortgage in July; my contract began in August; and by November or early December, the university had announced that it would close permanently.

What did you do then?!

I took a job as a researcher in the Faculty of Medicine at the University of Alberta. Then, 10 months later?

The research group closed.

Right about that time (May 2010), I realized how useful it is to live in a city that is packed with colleges and universities. Another local university gave me a 1-year full-time sessional teaching contract (extended to 2 years), complete with prescription drug coverage, which was very important to me because also in May 2010 I was diagnosed with Crohn's disease. The unplanned second year of my contract covered a sick leave for a faculty member. The pieces clicked together: people with continuing contracts have more options. By then, my condition had gotten quite bad, and it was interfering with my ability to be an effective university teacher.

So I laid out all my options, made several lists and even more phone calls, and decided to leave academia. I convinced the department where I did my PhD to hire me in a continuing, nonacademic role, which I did efficiently and I did well. I'm proud of the work that I did in the role, and I was incredibly grateful to the department for the steady employment and the excellent health insurance. Moreover, the flexibility the department gave me enabled me to more effectively manage my health conditionsymptoms effectively. I'd thought my exit from academia was permanent, too, until my friend Dan convinced me to—well, to move to Vancouver, actually. The idea to return to academia as part of that move was mostly mine. I think. My health was better and no longer driving the narrative, I was back to supervising undergraduate research and teaching classes, and I missed being a professor. I really missed it.

How have your career plans changed as you've continued on to your current position?

For the first 4 years of my current position, my career plan centered on stability and brevity. I aimed to put myself in as strong a position as possible to be awarded tenure, and I designed a "cut-and-run" strategy geared toward saving enough to either retire or shift to half-time work around age 50. But then, in my first year of eligibility, research methods students nominated me for one of the university's two annual teaching awards—and now that framed award recognizing innovative teaching has pride of place on my office wall. A year or so later, I had a particularly productive year, research-wise. Grant-funded undergraduate researchers made substantial contributions to that work. And then, a few weeks after being awarded tenure, one more thing happened: my dean shared that he values the contributions I make to the department, noting, "I don't see you as a(nother) yellow pencil. I see you as something else." These three experiences, taken together, led to a revised strategy. Rather than shuffling toward early retirement, I am thinking that I would like to aim to be in a position to apply for promotion to full professor as soon as I meet the length-of-service requirement. I would also like to explore whether there might be something else, in addition to my regular classroom teaching and research supervision, that I can uniquely offer my program, or the university. And rather than retiring at age 50, maybe that's when I might like to find out what it's like to spend the afternoon reading papers in a coffee shop.

If someone currently finishing their PhD was considering a similar position as you have now, how might they decide if it would be a good fit?

Find a mentor, someone who loves to teach and who is good at it. My teaching mentor was Connie Varnhagen, a 3M award-winning teacher of psychology. Even though most of what she taught me about showing respect for students, finding noncoercive ways to encourage participation, and plain having fun in the classroom went completely over my head at the time, over time those lessons have settled in and become a core part of my own teaching practice.

Also find out whether you like to teach, how often, and whether you are willing to teach to department need. The particular department need when I was hired was methods, so I offer at least three sections of junior and advanced level methods courses every year. Luckily, teaching methods has turned out to be one of the most satisfying aspects of my job. I really enjoy the other courses that I teach, too. But in 5 years, I have only taught health psychology, my area of specialization, once. The reviews were very positive, but when it's a choice between offering a core course (say, social psychology) and a specialty course (health psychology), the core course has to win out.

Another factor to consider to help you decide if a position like mine is a good fit is to find out whether there are options for faculty to adjust their teaching and research commitments; at my university, for example, the teaching load ranges from four courses to eight courses.

Next, ask yourself whether you can do research on a shoestring budget. Make a list of the minimum resources that you would need the university to provide (e.g., library resources, lab space, a budget for equipment), and think creatively about what you might study if the university you end up at cannot offer those resources. At my institution, there is a strong expectation of at least a moderate level of productivity, and this is true of many (though not all) teaching-intensive institutions. Research matters for annual evaluation, for promotion, for tenure decisions, and to maintain your reputation among your peers on campus and in the broader academic community. It could be in some cases that you can do minimal research and keep your job, but it will delay your own career progress and might close you off from certain opportunities. A position at a teaching-focused university may not be a good fit if you prefer to spend most of your time doing research then, but it also might not be a good fit if you have little interest in maintaining an active program of research, or little interest in involving undergraduate students in your work. Mentorship and exposure to experiential learning opportunities is a key feature of teaching-focused universities.

Another feature of my position is that it is located at a religious (Christian) university. Some religious universities will want you to acknowledge that you respect their faith tradition. Others will want you to share that tradition, and still others will want you to endorse very particular denomination-specific beliefs. Find out which it is before you prepare your application package.

More broadly, watch and listen for clues at the interview that the institution and the department will be a good fit. Ask if you can sit down with a prospective colleague, and listen—really listen—to what they say to you and how they respond to your questions. How do the faculty interact with each other and with administrators? Who says what outside of the formal interview periods, and how do others respond? Once, at a 2-day campus visit, I knew before sitting down to breakfast on the first day that I would not accept the position if it was offered to me.

Are there some aspects of running a research lab with undergraduate volunteers that someone currently finishing a PhD—in a lab with all of the related infrastructure—might need to consider? Any advice for them?

Someone running a lab with all undergraduate volunteers likely will not have much in the way of infrastructure. I found out recently that students in our program refer to the shared psychology lab space as a closet. It has a countertop desk, two desktop computers, three old chairs, two small filing cabinets, a tripod that I donated, and little else. Three faculty members share the space. This situation isn't too unusual for a primarily undergraduate university. But there are ways to overcome the limitations! Project management happens on Evernote, materials are posted to cloud storage, a small collection of "how to be a researcher" books live on a shelf in my office, and papers are downloaded to Zotero. We collect most data online.

Now, advice. A lab manual will help you and your volunteers in so many ways. Write a sentence or two articulating the lab's purpose. A statement of purpose can give context to, for example, expectations regarding how each lab member should treat colleagues and participants. If you can, illustrate the connections between your lab's statement of purpose and how you do research. For example, in my lab, any project pertaining to disability or involving marginalized groups must honor the "nothing about us without us" principle that comes directly from disability advocates. Authorship guidelines state the conditions under which volunteers will be included as authors, or thanked in the acknowledgments section, on published papers or in conference presentations. Include a passwords and procedures section to keep key information together in one place and easy to locate. Also state, as clearly as you can, what you can offer volunteers. Will you engage in collaborative decision-making? Will you write reference letters? Do you buy coffee? Are lab members given priority for honors thesis supervision? The appendices to our lab manual include sample research materials, including consent and debriefing forms, as well as common demographics (plus lab policies regarding when to collect particular types of demographic information and how to frame each question).

The other key piece of advice that I can offer is to actively pursue strategies that will accomplish the joint goals of (1) helping your undergraduate volunteers grow their skills and (2) saving you time. Some examples:

- Encourage collaboration on each other's projects. Each project will be better for drawing on the unique knowledge, skills, and experiences that each lab member has to offer. Working collaboratively also normalizes sending work to each other for feedback. Too many graduate students express insecurity about sharing ideas or showing their writing to others, but these are key aspects of scholarship. Make it supportive, keep it matter of fact—and if someone makes a substantial contribution to a colleague's project, consider formally adding them to the project team. This collaborative approach is how my current most senior lab member wound up with two papers in print before graduation.
- Offer volunteers opportunities for leadership and increasing independence over the course of their time with the lab.
- Do everything you can to help turn your volunteers into paid staff. Many students have scant time to do volunteer work when there are bills to pay. Our lab's recent increase in productivity is directly connected to the summer research assistantship program recently introduced at my university. And both lab members who won the grant last summer have landed funded positions in terrific graduate programs.
- Try to get papers out the door before your co-authors graduate. Life will get even busier following graduation, other priorities will claim their time, and they can

wind up being hard to reach. If you plan to publish *with or without* the involvement of your lab members following graduation, make that clear too, and address whether changes in their level of involvement might affect their author status.

If someone was interested in pursuing a similar career path, what would you suggest they do to better prepare themselves?

Sign up for any and all teaching workshops, certificates, and mentorship opportunities offered by your university. Then, gain some teaching experience at your university or at a local college.

Start developing a teaching philosophy. What would you like students to take away from each course? How will you design your courses to reflect those goals? Who are you and who do you want to be as a university teacher?

Build collaborations with others. Can you share resources? For example, my friend at a major research university used her research pool access to collect the data for our most recent study, and in return, I am using my professional development funds to cover most of the other costs associated with the project.

What do you like most about your work?

At this point in my career, what I like most is feeling deeply satisfied with the positive, productive relationships I have built with students, colleagues, and others. I also greatly value the independence and flexibility of my position, feeling like I have a voice at the university, and being granted the time and space needed to set and meet long-range goals. I have a real sense of control over my own career. It's an empowering experience.

What do you like least about your work?

The primary challenge is the difficulty involved in balancing the huge number of tasks that need time and attention each week: class meetings, lecture prep, email, marking, office hours, connecting with colleagues, data collection, writing manuscripts, revising manuscripts, undergraduate thesis supervision, student advising, committee meetings, meeting with collaborators, paperwork, department planning, grant applications, writing reference letters, and faculty meetings. My dream job is exactly the job that I have now, just 20% less of it. I rely on Evernote and a white board—and, let's be honest, my TAs and RAs—to help me keep track of everything that's going on, but I have not solved one recurrent problem: failing to leave myself enough time each week to work on the sorts of tasks that require a few hours of uninterrupted thought. Moreover, although I rarely miss deadlines or forget to attend meetings, I constantly have this sense that I am on the edge of failing to do something important. I have dreams about forgetting to go to class or showing up for class with nothing prepared. The shift to working from home and teaching online during the COVID-19 pandemic has only increased that sense of not being able to stay on top of important tasks, so I have arranged to have 20-minute daily meetings with a personal assistant. They keep track of deadlines small and large and help me double-check that I have dotted my i's and jotted my t's.

A second challenge is not having a clear sense of what is "enough" or what is "good." Is one paper in a year ok? Are four papers in a year good, average, or satisfactory? Should I present at more conferences, attend more campus events, take on more undergraduate research mentees, and apply for more grants? What happens when the answer to each of those questions is yes?

Based on your journey, what is some advice or suggestions you would want to pass on to someone who's currently finishing their PhD?

Start working with your campus career center to plan for what's next. Use their resources and other resources to help you prepare for the academic job market, for example, Inger Mewburn's *The Thesis Whisperer* and Karen Kelsky's *The Professor Is In*. Work with someone who has a strong knowledge of what makes for a solid letter of intent, teaching statement and portfolio, and other application materials. Ask that person to help you anticipate interview questions and prepare responses to each of those questions, including illegal questions (they are uncomfortable and they happen) and awkward moments (ditto). If you might be asked to give a research talk or to teach a class, arrange for knowledgeable others to give you feedback on content and, especially, delivery. You cannot over-prepare for an academic job interview, but you *can* be ready to handle, with aplomb, anything that is thrown at you.

Tailor each letter of intent to the specific institution and also to the type(s) of institution that you want to work at. An outstanding candidate whose letter reads as though they are much more interested in full-time research than in prioritizing their teaching may not be shortlisted for an opportunity at a teaching-focused, primarily undergraduate university, for example. If you are applying for a position at a community college, or a liberal arts university, or a polytechnic, make it plain that you understand and share the mission of that type of institution. I tell my students who apply to graduate school to show rather than say and to be specific rather than general, and the same advice applies here. Show the hiring committee that you want to be their colleague and that you are invested in building your career at their institution.

In your letter of intent, articulate what sets you apart from other job candidates. Be motivated, be precise, and use examples as evidence. For example, if your goal is for students to translate knowledge of the course content into action, explain why this goal matters, and then tell the committee exactly how you will meet that goal each semester. In the letter of intent for my current job, I described student posters on display at nonprofit organizations, as well as a public talk and colorful brochure created by students for family and visitors at a local long-term care facility.

Don't buy into the myth that your value is tied up in being a scholar.

As much as possible, always aim to have a plan B. Others have suggested that my, ahem, extreme dedication to having a backup plan at all times stems from my working class background. I'm not sure, but I do know that it does more than help me sleep at night: it also helps me quickly rebound after a career setback. My backup plan when I applied to graduate school in psychology was library school. My backup plan for a career in academia was to pursue a second PhD, in epidemiology. My backup plan at any given time for an unexpected job loss is a different type of academic *or* nonacademic role at a university. I have lost two academic jobs due to closures, but I have never been without a contract for full-time work.

Everyone reading this book recognizes the current state of the academic job market. It can take years to land a tenure track position. If you do get on the tenure track, your new department, faculty, or university might not be a good fit. University and research group closures are uncommon but not unheard of. It could be that you might decide to consider alt-ac (alternative academic) and nonacademic options, not only as a stop-gap measure but as another option for a rewarding, fulfilling career. You will find myriad ways—big and small, obvious and creative—to bring your skills to bear in whatever career path(s) you choose. I enjoyed my nonacademic position a great deal, and I would do it again.

You may also decide to consider pursuing short-term contracts to help you through leaner times: writing textbook ancillaries and sessional teaching, for example. Sessional teaching is important work (I still do it, and I love it). It can be deeply rewarding, and it can also help you get, or keep, a foot in the door at an institution. However, relying on sessional teaching can be stressful, even grinding, in the long term: teaching four classes at four different institutions is exhausting, the pay is usually low, benefits are often nonexistent, and courses can be cancelled at the last minute. In some cases, a department's current sessional instructors are no more likely than are other applicants to be shortlisted for a job opening.

For as long as you remain active on the academic job market, though, do whatever you can to stay involved in research at some level. Recently, I asked a colleague why the hiring committee had selected me over other candidates back in 2015, given that I had been working in a support staff position. They replied that the clincher was my continued involvement in research and research mentorship.

Decide what matters most to you: where you live, lifestyle, degree of independence in your work, using your PhD training, job role, long-term goals. Center those values in your career-related decisions. It's ok (awesome, really) to choose being a good family member over a job offer, for example—but find out whether it might be possible to have it all.

Is there anything else you'd like to tell someone reading this interview?

I have a tip: Find out which courses nobody wants to teach, and convince yourself that *you* do. If you can, teach those courses once or twice before you go on the job market. Use that teaching experience as evidence of your commitment to the course that you are fairly certain the second most recent hire can't wait to hand off to you.

Thanks so much for sharing your experiences with us, Kelly. It is much appreciated!

"Be Bold and Go for Opportunities Outside Your Comfort Zone"

Elliot A. Ludvig

Elliot A. Ludvig

Abstract In this chapter, Elliot Ludvig tells us about his journey through several departments—psychology, neuroscience, computing science, and mechanical/aerospace engineering—as well as to industry and back again. Elliot shares how his goals led to this path and suggests that readers should be bold and going for opportunities outside their comfort zone. Research responsibilities change with career progression, but other roles become important in academia as well, including teaching and mentorship, with the potential for others, such as being a course director. Elliot has also moved internationally and discusses some comparisons between countries. Other challenges can also be critical, such as the 'two-body problem' of securing jobs for both partners and balancing academia with having a family life.

Contents

E. A. Ludvig (✉)
Department of Psychology, University of Warwick, Coventry, UK
e-mail: E.Ludvig@warwick.ac.uk

C. R. Madan (ed.), *Academia and the World Beyond*,
https://doi.org/10.1007/978-3-030-82606-2_19

Chris: Can you introduce yourself and tell me a bit about your current position?

Elliot: I am a behavioural scientist who studies how people and other animals learn to make effective decisions. My research draws on ideas from a few different fields, including computational neuroscience, artificial intelligence, and comparative cognition, which I learned through multiple post-doc positions.

In that peripatetic phase of my career, I did three post-docs that were housed in three different departments (Neuroscience, Computing Science, and Mechanical/Aerospace Engineering), none of which matched my initial degree in Psychology. I also spent a year and a half working in industry at a drug discovery company. Despite the variety, the questions I asked in each position were often highly related, and I learned new tools and techniques for answering these questions with each opportunity.

Currently, I am a professor in Psychology at the University of Warwick in the UK. I lead a research group with two post-docs, three to five PhD students, and typically four to six MSc students. My lab is part of a larger conglomerate at Warwick that we call the Warwick Modeling Behavior Lab (WMBL, with members affectionately known as Wombles). I am also the course director for the MSc in Behavioural and Economic Science (BES) at Warwick, which attracts upwards of 50 amazing students from around the world every year to study in a course that spans psychology, economics, and the business school.

What was the focus of your PhD?

I finished my PhD in 2003 at Duke University in the Department of Psychological & Brain Science (changed partway through from Psychology: Experimental). My PhD research focussed on timing in pigeons—it was mostly an empirical PhD project, and I spent many an hour carrying the birds around the lab (in plastic pitchers!) and waiting for them to finish up their sessions for the day.

In that project, the main thing that I was examining was how pigeons learned to adjust their behaviour when the time between rewards changed rapidly in their environments. Most research in the timing field looks at steady-state behaviour—what happens after animals have been trained for weeks or months on end with the same exact time relationship between cue and reward. My project instead looked at the question dynamically: what happens if the time between cue and reward varies on every trial, but in systematic ways? It turns out that pigeons could learn to adjust their timing really quickly—something that most theories cannot handle. Pigeons could even learn to anticipate future changes if the patterns were simple enough.

One of the best things about my PhD, though, was that I got to take graduate-level classes across the university. Outside psychology, I took classes in philosophy, physics, literature, and computer science—all of which helped inform my thinking about my core research questions. It was also in those classes that I really got a taste of what can be learned from other disciplines, setting the stage for my discipline-jumping post-doctoral years.

As you were finishing your PhD, what were you thinking about your career plans?

Back when I was finishing my PhD, my clear long-term goal was to secure an academic position. I had a post-doc in neuroscience lined up, and my goal was to spend 2–3 years in the position, learn some new techniques, publish some quality papers, and land an academic position at a research-intensive institution. At that point, I had had such a positive experience in the last few years of my PhD (after a rough start, but that's a story for another day) that I was convinced that an academic career was the only way to go. In particular, I was enamoured with the intellectual freedom that I had been afforded in my PhD–given free rein to study any question of interest and read/write about whatever I found to be compelling.

People do a post-doc in an adjacent field often enough, but doing several, each in different fields, is less common. Can you tell us more about how you ended up progressing along that path?

The jumping around was not wholly planned. As is often the case, it was one part opportunity and one part necessity.

The shifting between fields represented to some degree an intellectual restlessness. With each post-doc opportunity, I was learning a new skill set, but as I learned more, I also felt that there was something missing in my intellectual toolkit. So, for example, in my first post-doc (at Rutgers) and in my industry post (at PsychoGenics), I spent a lot of time collecting data with rats and mice in various learning experiments and other cognitive tasks. As I wrote these into papers and reports, despite the experimental rigor of the designs, I felt that there was a lack of theoretical coherence to the work. It felt like a bunch of disconnected facts. This discomfort led me to seek out an opportunity to focus solely on theory, and I was fortunate enough to secure an opportunity in a computing science department (at Alberta) in their reinforcement learning group. A similar saga with a desire to focus more on neuroeconomics and human decision-making led me to Princeton and my final post-doc (in Mechanical and Aerospace Engineering and the Princeton Neuroscience Institute).

On the necessity front, the multiple post-doc sequence was driven by my desire to secure an ever-elusive research position in academia. Along this journey, I applied to countless faculty positions; had many phone, skype, and on-campus interviews; and even received a couple of job offers. But none yielded a suitable permanent position, so I guess it was a matter of stubborn persistence. The post-doc opportunities were (fortunately) there, and I was continually learning new things and immersed in constructive and engaging intellectual environments, so it was almost natural to keep going until eventually I found an academic post.

There was also a personal angle. I was always trying to resolve a two-body problem with my wife, who is also an academic, but in a very different field (literature). Each move represented an attempt to satisfice across intellectual interest, career progression, a steady paycheck, and appropriate opportunities for my wife. For example, my move to Alberta was in part necessitated (or enabled, pending the perspective) because the college that my wife was teaching at closed down—and we needed to find something new.

Can you tell us more about your time working at the drug discovery company? For instance, what did the job involve and some of the considerations that led you back to academia.

At the drug discovery company, I was the research lead for their cognition group. The company had around 50 employees at the time, and I led a small group of around four to six research scientists. My main role was to develop and implement cognitive tasks with mice (e.g. spatial memory, timing, or associative learning). These were often challenging tasks for the mice where performance could be improved by nootropics (cognition-enhancing drugs) or where there were severe deficits in particular genetically modified mice. The goal was to find novel drugs that could prevent or reverse cognitive decline. We did both in-house work (i.e. trying to find our own drug) and contract work for big drug companies, where they would send us some unspecified compounds (only given a number) and ask us to assess which drugs held the most promise.

The work was challenging and intellectually engaging. I particularly liked the interface between science and business—something that I had not been exposed to at all in my graduate training. I even was able to publish a fair number of our findings, whenever we developed a new behavioural test or learned something new about a well-known compound or neurotransmitter system.

After a while, however, what I found was that there was a limit to how far I would be able to go with only limited academic experience and publication record. In all the companies with which I interacted, the most interesting positions to me (e.g. the chief scientific officer or head of R & D) were always held by people who had spent a lot longer in academia and had built up a stronger publication record and research reputation before embarking on their industry careers. So I felt (perhaps mistakenly) that, even on this industry track, I would need more academic experience to get to where I might have wanted to be.

When PhD students think about a future in an academic position, I don't think they often consider roles such as being a course director. What does that entail and how does that fit with your other responsibilities?

Being a course director is one of the most fulfilling elements of my job as an academic. At Warwick, I am in charge of a master's degree (course) in Behavioural and Economic Science. In that role, amongst other things, I help design and adjust the curriculum, create a sustainable budget, recruit prospective students, and develop partnerships with industry. Because the degree spans psychology (my department), economics, and the business school, I also spend a lot of time interacting with academics in different departments to coordinate our teaching approaches and materials. It's a role that requires lots of different skills.

The best part though is the students. Around 50 students from around the world attend the course every year. We have had students from every continent and over 30 countries, just in the last few years, so it's a fascinating group of students with whom to interact, and I also stay in touch with many of them after the course ends. Many have stayed on at Warwick to do PhDs after their degree, and I even serve on the scientific advisory board for a behavioural science company that was founded by two of our alumni.

What are some of the different skills that you think your role, or academic positions more generally, require?

An academic position really requires people to be good at lots of different things. There is surprisingly little specialization, and academics need to excel at many different facets of their job from the usual research and teaching, but also people, time, and budget management.

I'll highlight three skills here that in some ways represent three different ways that people can be successful in academic careers. The first is technical competence—in whatever your chosen field or sub-area, you need to be able to execute the core research to a high standard, be it experimental work or theoretical work. This is especially important in your early years when you are executing most of your research yourself and as you move on to have your own lab and train others in the skills you have gained.

Over time, however, I think this technical skill becomes less key as your lab members become the technical experts, and your role shifts more to the other two skills—communication and strategic thinking. Being a strong writer and speaker is always important. It helps with getting your research noticed, understood, and funded. It helps with teaching. And it helps with getting things done in a larger institution, like a university. A particular aspect of writing that becomes really important is the ability to edit and provide effective feedback—I find that is perhaps the largest part of my role these days, both in research and teaching.

And the final skill is strategic thinking—being able to take a step back from the minutiae of the current task and asking good big-picture questions about what is important to achieve and how to best go about doing so. This sort of thinking is crucial not only in identifying and usefully pursuing research questions but also in taking on institutional responsibilities, like being a course director as I discussed above.

You've moved around between countries through your career. How different did you find the academic systems?

I have spent time at universities in Canada, the USA, the UK, and Israel, so I have indeed seen a few different academic systems. In my experience, I think the two things that vary the most across systems are (1) the degree of specialization in undergraduate degrees and (2) the degree of flexibility offered to staff in terms of their teaching.

In North America (especially the USA), undergraduate degrees are very broad, following the liberal arts tradition. Students take classes in many different subjects, often before choosing their major, where they then will take a higher proportion of their classes in their chosen major. In the UK instead, students apply directly into a specific degree (not just to a university) and typically will take a prescribed set of classes to meet the requirements of the degree, with only limited optionality. The degree ends up much more focussed that way and usually shorter by a year, but I feel there's a substantial loss in terms of the education. Part of being in university is learning new things outside your comfort zone—and I feel that's less possible in the UK system. As with so many other things, Canada is a little more in the middle with a higher degree of specialization, but nothing like the UK system.

A second difference is in terms of the flexibility and autonomy given to academics. In the UK system, there is a lot more oversight. Things like the assessments in a class (e.g. essays and tests, etc.) need to be decided far in advance, sometimes 18 months or so. Exam questions often need to be checked by an external examiner from outside the university before a class is even run. Marks are moderated, sometimes second marked, and then cross-checked by an exam board before being released to students.

This oversight helps with making sure that things are well prepared and consistent across classes and even across universities. The sort of scandals that occasionally pop in the USA where athletes are given preferential treatment in fake classes would be near impossible in the UK. On the other hand, this rigid pre-planning can really restrict the range of teaching approaches and even the content that you can present to the students. There is a lot less of the classroom autonomy that one has in North America in terms of creating the structure of the class. As I do most of my teaching at the master's level—where there are fewer restrictions—I have managed to avoid many of the pitfalls of the system, but still making on-the-fly adjustments to improve the learning experience is very difficult.

Can you tell us a bit about what day-to-day life is like in your current position?

As with most academic positions, day-to-day life varies quite a bit pending the time of year. In term time, my time is dominated by teaching responsibilities (e.g. preparing lectures, seminars, and assessment materials, delivering the classes, answering student emails, and marking). Outside of term time, research plays a much more prominent role, and I spend lots of time talking to collaborators and my research students (PhD and MSc students).

More broadly, I would say that my current position is very much dominated by providing feedback on materials created by others. Many facets of my job require this—from manuscripts, grants, or pre-registrations written by colleagues, to student papers, to manuscripts for review, to new experiments that we will run in the lab. A big part of the job is reading or watching and then providing as constructive feedback as possible to make the work better.

In day-to-day life, this translates into half days of reading materials and providing feedback and the other half day with a string of meetings about research—either with students (PhD or master's) or collaborators. For a long time, I have tried to carve out 2 days a week with no meetings scheduled so that I can focus on deep work, like writing or coding. This unfortunately has become more challenging over time as my responsibilities have increased, and I often find now that a month goes by without such a meeting-free day.

What do you like most about your work?

The freedom to research whatever I find interesting. That is definitely the best part of an academic career—there really are few (no?) constraints on what I can spend my time reading and researching. In practice, I am embedded in a social network, and a lot of what I end up doing is influenced by my students, colleagues, and collaborators, but, at heart, I get to spend my days studying the problems that I find most interesting and important, using the tools and techniques that I think are most effective. It's a luxury and privilege to have that as my work.

The other really amazing aspect is the people with whom I get to interact on a regular basis. On the one hand, I get to teach and mentor really smart, motivated, and interesting students, who then go on to do cool, impactful work around the world. Staying in touch with the students and learning how they apply what they learned about behavioural science in industry, in government, and/or in their future research is really a source of pride. At the same time, I get to work with outstanding colleagues and collaborators, from whom I learn a tremendous amount on a regular basis.

And what do you like least about your work?

Also the freedom. The ability to spend my time however I think most fruitful research-wise is an incredible boon, but it's also very challenging and occasionally disheartening. First, there are so many interesting problems out in the world to study, so narrowing down to an achievable chunk really takes substantial effort. There is often the distressing feeling of 'What if I'm wrong? What if I am wasting my time on this project/paper/technique?'. Second, because I am deciding what to work on and what's important, when the world sends back rejections (for grants, for papers, etc.), that is doubly disheartening. Of course, I have developed a relatively thick skin about this sort of thing with time, but the rejections also introduce the nagging doubt that my work may not be worthwhile.

If someone currently finishing their PhD was considering a position similar to yours, how might they decide if it would be a good fit?

I think one of the things to consider is whether you would prefer to continue to do hands-on research like you do in a PhD or if you would rather be one step removed and manage/teach the research instead. In my position as a mid-career academic, I very rarely do the hands-on research anymore: coding up the experiments, collecting the data, running analyses or simulations. Sure, I spent much (most?) of my time doing research, but that is in different roles. I write grants, help develop ideas for new experiments and models, provide feedback on manuscripts and experiments, present the lab's research at conferences and colloquia, review grants and papers, help edit journals, and organize conferences. So if you like the real hands-on element of research, then looking for a research scientist post might be more suitable than the standard academic career.

A second thing to consider is how much you like to teach; a good portion of the academic role is teaching. Not only classroom teaching but also mentoring research students of all levels and higher-order teaching items like designing a curriculum (more fun) or a new plagiarism policy (less so). If you don't like teaching, then an academic post (even a research-heavy one) is probably not suitable.

If someone was interested in pursuing a similar career path, what would you suggest they do to better prepare themselves?

1. At first, I'd say the same thing to anyone tempted into this career path: Don't do it. Seriously, don't pursue an academic career. Look for a job in industry or government or self-employment where your skills will be more valued, better remunerated, and have a greater impact on the world.
2. Now that I scared away everyone with that first point, let me follow up with an allegory. If someone wants to convert to Judaism, the Rabbis always tell them 'no' the first time. And the second time. Only if they come back a third time, showing a real passion and diligence, are they offered the chance to spend many years studying for the right to convert. I find this a very apt analogy to an academic career—you should really only consider it if you really cannot stay away. That might mean trying something different for a while (like I did with my industry post-doc).
3. To those who persist, my next piece of advice would be to prioritize quality over quantity in your research. Take the time to do high-quality research. Bundle your best work together into a handful of really impactful publications, instead of salami slicing the work into more, smaller papers.
4. Get teaching experience. Find a way to get in front of a classroom with an opportunity to teach the students something new. You don't need a lot of teaching experience, but it really helps to have some in terms of both securing a position and making sure that the career path is right for you.
5. Choose your research questions wisely. It helps to read deeply—don't just skim abstracts of lots of papers to know what's out there, but rather really dive in and gain some mastery of the material. That will really help get your work to the edge of the field. And then get lots of feedback. Present your work whenever you can (at conferences, at lab meetings, in other labs, in local seminars) and use the feedback to refine your thinking.

Earlier you mentioned that your wife is also an academic and the issue of the two-body problem and both securing jobs. Can you tell us about some of the considerations that came up when you were both interviewing and trying to find positions or other related advice?

The two-body problem has really shaped my career to some degree. With each possible job that I have considered, one additional consideration has been what my wife would do—and vice versa, whenever she has considered jobs. The repeated moving that characterized my early career made this a point of continuous negotiation.

What helped us through this issue was being very flexible in terms of next steps—we didn't limit ourselves to one type of job or one specific employer/location. We applied broadly and repeatedly, always trying to line up the next position before the current one came close to expiring. With each opportunity, we had to consider the usual things about the job (e.g. career progression, research opportunity, teaching prospects, location, salary, etc.) but also had to consider the likelihood of finding a good job for the spouse (e.g. number of nearby universities, types of research in related departments, other non-academic options). We turned down multiple (tenure-track) opportunities along the way, where it did not seem that the other spouse would have a suitable opportunity nearby. In the end, we were very lucky and the 'trailing' spouse (which was my wife for three out of four moves) always managed to find a suitable opportunity within a few months of arriving in a new location.

I know you both also have children. Do you have any advice for those thinking about when to have children or on how the associated change in lifestyle relates to life in academia?

We have three kids, and I had all three kids during my post-doctoral years. Obviously, having kids as a post-doc can be a serious financial strain (childcare costs can approach your salary). In addition, they are remarkably time-consuming little creatures, so the amount of time and mental space available for progressing your research career definitely takes a hit. There is never a 'good time' to have kids—wherever you are in your career, kids will certainly be disruptive.

On the other hand, I felt that I gained a degree of focus with the kids around. Knowing that I only had a smaller number of hours to get things done helped make those hours much more focussed and on task. When my wife was pregnant with our first child, she was working on her dissertation in literature, and the prospect of a newborn propelled her to finish and defend while 8 months pregnant. Those time constraints can really serve as a good impetus to keep you focussed and on track.

In some ways, academia can be very well suited to having children. The hours are typically very flexible, and working from home (even before the pandemic) is often possible. I was fortunate enough to have generous and sensitive supervisors while the kids were very little, so that made things easier as there were limited hard constraints on the exact timing and location of my work. This made the work-kids juggle more manageable. My department now is also very family friendly—important meetings are only scheduled during core hours (i.e. when kids are in school), but there is wide variety across departments, disciplines, and countries, in terms of the norms.

I think the big challenges emerge from the financial and geographic uncertainty that can come early in an academic career. Kids are expensive, and postdoc salaries are often low enough that making ends meet can be an extra challenge with more mouths to feed. Added to that is the positional/geographic uncertainty. While in a precarious academic position, you might have to move on short notice, and kids take planning in terms of care arrangements and schooling. Plus, they have friends of their own, so the social challenge of (repeated) displacement can be tough on them.

One last, important thing to note: I think having kids has actually made me a much better mentor and supervisor. There is a certain type of concern and patience that you develop in raising kids that I find transfers well into a work environment and makes you well placed to better support your students and other supervisees to help them achieve their goals.

Based on your journey, what advice or suggestions do you want to pass on to someone who's currently finishing their PhD?

I think I'd give two pieces of advice to someone finishing a PhD. First, I'd recommend that they take a step back and take stock of the skills that they have amassed during the degree. They might not realize it, but they have become very skilled in project management, communication, (often) computer programming, research design, and data analysis, amongst many others. These are very transferable skills that will be useful outside the scope of their current project.

And then do something (very) different next. Move into a lab in a new field. Look for an industry job that values some of those skills you identified above. Straying far from what you did in your PhD will help you develop a unique perspective and also help you refine what it is that you want to do for the next 10–20 years and beyond.

Is there anything else you'd like to tell someone reading this interview?

No career decisions are irreversible, so be bold and go for opportunities outside your comfort zone. You'll build up a unique profile and experience base that will help you chart your own path forward that will not end up looking like the careers of any of your mentors. The world is changing, and much of the best research is now happening outside academia, so there are lots of ways to put that PhD to good use.

Thank you so much for the thoughtful advice, Elliot!

“I Get to Travel a Lot and Talk to Many Really Smart and Thoughtful People”

Eiko I. Fried

Eiko I. Fried

Abstract In the interview with Eiko Fried, we discuss his career so far. Eiko shares with us some of the insights he has gained into how academia and expectations differ between countries. We also discuss the myriad of skills that are ‘useful’ in academia but cannot be mastered by a single individual, thus providing some motivation towards working in large-scale collaborations with a range of experts. Academia has many benefits including talking with brilliant people, analysing data, traveling, and teaching—but it can be difficult to treat academia as ‘just a job’ given its high workload and difficulty in securing a permanent position. Finally, remember to be kind, despite the international nature of academia; it can also seem smaller than you think.

Contents

E. I. Fried (✉)
Department Clinical Psychology, Leiden University, Leiden, The Netherlands
e-mail: e.i.fried@fsw.leidenuniv.nl

C. R. Madan (ed.), *Academia and the World Beyond*,
https://doi.org/10.1007/978-3-030-82606-2_20

Chris: Can you introduce yourself and tell me a bit about your current position?

Eiko: Hi! I'm 38 years old, was born in Germany, have lived in Germany (mainly in Munich and Berlin), Belgium (2 years), the Netherlands (4 years), and the USA (1 year, Michigan), and spent about a semester each in Norway, Finland, and the USA (Virginia). I am currently Assistant Professor in the Clinical Psychology Unit at Leiden University in the Netherlands. I simultaneously applied for a position here in the methodology group because my work takes place somewhere in between these two fields.

What was the focus of your PhD?

In Germany, unlike the USA, PhDs usually follow after bachelor and master's programmes, and will commonly last 3–4 years. After finishing what was equivalent to bachelor and master's in Munich (called a Diploma), I spent a year in Berlin at the Cluster of Excellence 'Languages of Emotions' as a research assistant because I had failed to get into the graduate school of that cluster. I got in the year after (you needed to apply with your own project proposal and then were assigned supervisors from the cluster after you were accepted) and took about 3.5 years to finish my PhD there, 2011–2014. I spent some of that time at the University of Michigan in Ann Arbor. The topic was covert heterogeneity of major depression: I studied all the ways people with depression differ from each other and showcased how studying individual symptoms on which people differ a lot—rather than the yes/no category major depression that obfuscates a lot of information—can be highly insightful. At the end of graduate school, I had one paper published, one accepted, and one submitted. I want to mention this because I saw a discussion a few days ago on Twitter about US students in my field having on average 17 papers when they finish grad school (I did not verify this, but my CV was far from that, in any case).

As you were finishing your PhD, what were you thinking about your career plans?

I wanted to stay in academia, but I'm not a good planner and took things as they came. I had become a big fan of the work of Denny Borsboom in Amsterdam, and so before finishing up my PhD, I reached out to him and asked if there may be any opportunity for some future work together. Denny had heard about my work through a mailing list we were both active on (shoutout to SEMnet) and invited me to come to Amsterdam for a few days and present my work. I then ended up applying to a postdoc position in Leuven (Belgium) that Denny told me about which he co-supervised, and to this day, I am not quite sure why they accepted me (the position was very stats heavy, and my CV didn't look great back then compared to other candidates).

How have your career plans changed as you've continued on to your current position?

Despite all its challenges, I greatly enjoy working in academia. I worked hard but was also quite lucky: after my first 2 years of postdoc, I ended up not getting a grant, which, because I did not get the grant, led to another 2-year postdoc position, this time in the Netherlands in the group of Denny. I moved, and when this second postdoc ended, I really enjoyed living in the Netherlands. So I applied for two jobs at the geographically closest university, Leiden University (pro tip: it wasn't a good idea to tell them that; you're supposed to say something about how much you like the university you apply for), and ended up getting one of them. I've been there for 2.5 years now, at a place that starts feeling like a real academic home.

A few years ago, you wrote a blog post about the list of skills/competencies that postdocs can be expected to have. Can you tell us a bit more about that?

I actually revisited that blog post a few days ago when I wrote a rejoinder to commentaries that had been published on a paper of mine. The idea of the paper was that we need to pay more attention to theory building and testing in my field (psychology) and many of the commentaries suggested training psychologists in math, modelling, and philosophy to achieve that. So I collated a list of expected academic skills, expertise, and services, from, for example, job postings, university promotion guidelines, grant guidelines, and workshops on career advancement in my area. The list is exceptionally long: to give you just one of the 11 items on the list, clinical psychologists are often academics, teachers, and practicing psychotherapists all at the same time. Adding math, modelling, and philosophy to our education does not seem feasible. Instead, I suggest that we should all learn a bit of these and instead sacrifice some content that is outdated; train *some* psychologists to become theorists (like in physics, economics, or biology where theoretical subdisciplines exist); and collaborate more with interdisciplinary experts.

Can you tell us about a few more items from that list? I have found myself referring students to that list from time to time to demonstrate the breadth of directions there are to develop skills in and to show that's unfeasible to become an expert in everything that is 'expected', so it would be great to hear more about it, as well as a link to the blog post.

Of course, here is the blog post (https://eiko-fried.com/are-we-asking-too-much-a-list-of-competencies-people-expect-me-to-have/), and here (https://doi.org/10.1080/1047840X.2020.1854011) the paper I was referring to for which I used the blog post as an inspiration (the relevant part is in the very last section). I'm glad you're using this to normalize things a little with your own students … it's indeed impossible to obtain expertise in all these areas.

Let me give you an example for a recent paper we wrote on student mental health under COVID-19 (https://doi.org/10.1177/21677026211017839), for which we queried students on smartphones multiple times per day for 2 weeks about their momentary experiences during the early stages of the pandemic. The paper requires considerable expertise for a broad range of substantive constructs we wanted to measure, ranging from depression and anxiety over well-being and anger all the way to loneliness and social isolation. In the best case, you've read books, theory papers, and systematic reviews on each of these constructs. You also need measurement expertise, especially in the context of ecological momentary assessment for which you can only ask participants a very limited set of questions (because you query them multiple times per day for many days in a row). You need technical and programming expertise to implement these questionnaires properly in the right software. You need at least some expertise in privacy and IT security in order to get this approved by your ethics committee, given that sensitive data are collected on smartphones. Then you need expertise in dynamic systems models for the analyses we did. And expertise in how to deal with missing observations in time-series data. In the review process, we received three detailed reviews, and reviewers raised questions across all of these domains. This example also shows that the answer to proper science in these areas has to be large-scale collaborations of a range of experts.

Can you tell us a bit about what day-to-day life is like in your current position?

My experience in the Netherlands is that working from home is often encouraged at universities. In my last 2.5 years as an assistant professor, I have worked from home about 2 days per week. What I do with my week depends heavily on my teaching schedule: there are times in which I pretty much only teach (last semester I was supervising 16 bachelor students and six master's students and was additionally teaching two courses). If my teaching load is lower, in a given week, I will do the following:

- Several meetings with PhD, master's, and bachelor students I supervise.
- One or two committee meetings in our department or faculty, e.g. for research or educational purposes.
- One or two meetings with folks from the Open Science Community Leiden and the Young Academy Leiden.
- Focused work on a manuscript, which usually means reading, analysing data, data visualization, and writing.
- Procrastinating on Twitter.
- Providing written feedback for manuscripts; this can be for friends, colleagues, co-authors, and of course also for folks I don't know via the peer-review system.
- Many calls and email exchanges with collaborators about ongoing and future projects.
- Trying to reserve 2 hours for learning something new, but that doesn't always work out.
- Some blogging (although this has become less frequent these days) and science communication/catching up with science news on Twitter.

What do you like most about your work?

I get to travel a lot and talk to many really smart and thoughtful people. It may sound a bit cliché, but it's incredibly rewarding for a person as curious as myself.

And what do you like least about your work?

I've never been good at administrative tasks that have nothing to do with teaching or science (i.e. what I consider to be my job). Luckily, I've had jobs so far where I spent the large majority of my time on science and teaching, so I haven't struggled with this too much.

You've moved around between quite a few countries. Have you noticed any interesting differences between the academic systems?

Absolutely. I'll give you two examples: the first is the meaning of academic positions. I am currently an assistant professor in the Netherlands, but here this is really just a postdoc with a higher teaching load, often with a short-term contract. It does not come with any start-up money or PhD students or having your own lab, in stark contrast to how this works in the USA. The second difference is what terms mean in the English language. I'll be overgeneralizing a little here to make the point, but good luck trying to give a Dutch student critical feedback on an assignment or thesis the way you'd do that in the USA: 'This is really great work, Jessie, but you may want to consider the possibility to perhaps work on the structure a bit.' Dutch students will consider this a 9.5/10 and not move a finger. In the USA (at least where I have taught), this is more of a 6/10. Such differences are also reflected in letters of recommendations I write (in English language) for students who want to work in the USA vs the Netherlands or the UK: using superlatives common in US letters would simply sound weird for the Dutch. A 'good student' would not be considered in the USA, but in the Netherlands, a 'good student' is, well, actually good. As someone who isn't an English native speaker, it took me quite a while to figure out these nuances.

Can you tell us more about how you ended up working on your PhD jointly between Berlin and Ann Arbor?

In 2010, I was lucky enough to receive a competitive PhD grant to work at the highly interdisciplinary cluster of excellence 'Languages of Emotion'. You applied with your own project, and if it got funded, you then looked for supervisors among the faculty. There wasn't really a strong match between my topics and the faculty, so I ended up writing an email to Randolph Nesse, a professor at the University of Michigan, asking whether he'd be willing to provide feedback on my PhD thesis proposal (he had done a lot of work on my topic of interest with a previous PhD student around 2005). His response ended up in my spam filter, and my life would have been very different had I not checked my spam filter a few days before it was emptied automatically. Randy was enthusiastic about my ideas and became my informal supervisor, and I ended up visiting Ann Arbor twice for about a semester each time.

If someone currently finishing their PhD was considering a position similar to yours, how might they decide if it would be a good fit?

Some of my friends have jobs that are just that: jobs. Time spent to earn money to then have a good life outside of work. In my personal experience, academia is not a good place for 'just-jobs' jobs because keeping up with an academic career requires considerable motivation and time investment. I don't want to normalize this or pretend that's great, but I'm answering the question for academic careers as they currently are, not as I'd want them to be. Given the high workload and volatility of academic careers, I think it's important that folks really enjoy the prospect of science and teaching. I often feel that what I am doing at university are activities I do primarily because I enjoy them: talking to brilliant people, analysing and visualizing data, traveling, teaching, learning, trying to find out what holds the universe together at its core! It can be a little much at times, but I truly enjoy all these things. Academia can be challenging, and it's not easy to get a permanent contract. It helps if you are curious and bring with you an innate passion for science and a bit of intolerance for uncertainty.

Based on your journey, what advice or suggestions do you want to pass on to someone who's currently finishing their PhD?

Countries have very different academic systems, and different disciplines tend to have different systems as well. A good example from my own career is that I was recently assigned to be the last (i.e. least important) author on a paper by a large consortium of epidemiologists because they order by importance of contribution. That turned out to be really convenient for me because in psychology, the last author is considered the second most prominent position, implying something akin to PI status (we do first author, last author, and then second, third, fourth, etc.). So I admit I struggle with broad general advice. Maybe this: academia is much smaller than you think. You will, unexpectedly, meet the same colleagues over and over again, even if you work in a very large field. People you criticize on Twitter will end up reviewing your work. People you talk about badly at a conference dinner will have their spouse sitting at the next table. The message here is not 'be careful what you say in public' (although that may be a good idea), and it is certainly not 'do not be critical' or 'do not criticize others': criticism is central to science. But it is important to be fair and balanced in your criticism, focussing on the content and not the person, and that's not always easy. Assume a cockup before you assume a conspiracy. Ask for clarification if you see something that looks fishy to you: most errors in academia are honest mistakes. And in a scientific debate, try to follow Anatol Rapoport's rules, as summarized by Daniel Dennett in his book *Intuition Pumps and Other Tools for Thinking*:

1. Re-express your target's position clearly, vividly, and fairly.
2. List any points of agreement with the target.
3. List what you have learned from your target.
4. After that, your criticism, no matter how harsh on the matter of substance, will likely lead to a much more constructive debate.

I have a comic by Joey Comeau in my living room to remind me of that. It says: 'I used my one wish to make myself smarter. Smart enough to wish I was more kind.'

Thanks so much for sharing your experiences with us. It is very much appreciated!

"Your Skills Are Valuable and You Likely Have Many Options After Your PhD"

Simine Vazire

Simine Vazire

Abstract In the interview with Simine Vazire, we discussed her early PhD experiences and the lessons she's learned along the way. Simine tells us that much of her role is meetings, particularly with trainees. She has also had a variety of journal-related roles, from reviewer to editor-in-chief, and describes the related decisions as part of the peer-review process. Simine's work to improve the field resulted in her co-founding the Society for the Improvement of Psychological Science (SIPS). Time management and mentoring are very much part of the job description at her career stage. Simine further provides a wealth of career advice for doing well in academia and shares her experience in moving between universities and internationally.

Contents

S. Vazire (✉)
Melbourne School of Psychological Science, University of Melbourne, Parkville, VIC, Australia
e-mail: simine.vazire@unimelb.edu.au

C. R. Madan (ed.), *Academia and the World Beyond*,
https://doi.org/10.1007/978-3-030-82606-2_21

Chris: Can you introduce yourself and tell me a bit about your current position?

Simine: I'm currently a professor in the psychology department, officially the "Melbourne School of Psychological Science," at the University of Melbourne. I moved there, from UC Davis, during the pandemic, so I actually don't yet have a physical office there and haven't actually moved to Melbourne yet – but I made it as far as Sydney (where my partner lives). My position there is a combination of research, teaching, and service. Within the department/school, I'm in the "Ethics and Wellbeing Hub," which is basically like the social/personality area. My teaching is focused on research methods and professional development. Most of the service I do right now is linked to journals and editing.

What was the focus of your PhD?

I did my PhD studies from 2001 to 2006 in the social/personality psychology area in the psychology department at the University of Texas at Austin, with Sam Gosling as my advisor. My research was mostly on self-knowledge, or the accuracy of self-reports of personality and behavior. Do people know what they're like? How others see them? How they behave? It involved a lot of measurement/assessment work, like collecting self-reports and informant reports from many different kinds of informants and trying to figure out the best way to measure actual behavior. Then I'd triangulate those measures and try to figure out which ones are more accurate. It was very methodological, but it also was trying to get at some thorny philosophical/theoretical issues, like how well people know themselves, why self-deception happens, when can we be wrong about ourselves and when are we by definition the best expert or authority on ourselves, etc. Most of the time I used concepts and methods more common to personality psychology than social psychology (e.g., the Big Five), but I thought of my work as being at the intersection of personality and social psychology.

As you were finishing your PhD, what were you thinking about your career plans?

I knew that I would prefer to stay in academia if I could, but I also didn't want to take a job that would involve doing a lot of things I don't enjoy doing. At the time I don't remember a lot of awareness of non-academic jobs, so that wasn't really on my radar. In my mind, I would try for an academic job, and if I couldn't find one I liked, I would probably just pursue a totally different career path (something where I could be self-employed, probably – I really hated the jobs I'd had where I had a boss and a timecard etc. I really value my freedom and independence). I remember having lunch with a very senior person in my field and he asked me "why do you want to be an academic?" He was very scary and notorious for having – and expressing – strong opinions, so I knew there was a right answer. I panicked and just told the truth that I wanted it for the lifestyle (to be able to work when I want, from where I want, and often on the topics I want). Turned out he thought that was the right answer. I still don't know if it's true that academia is unique in that way (maybe now it's more common in other industries too to have that kind of freedom), but academia has certainly lived up to this particular promise for me. I've had incredible freedom to have unconventional hours, travel and move around a lot, and just generally not have to conform to strict rules or have someone looking over my shoulder all the time. The main things I was dreading about academia were having to write grant proposals and maybe not being able to choose where I live, though I was also pretty open to living in a lot of different places. There were just a few specific places *I really* didn't want to live (College Station, Texas, and anywhere in/near Minnesota, specifically. I did college in Minnesota and it was too cold for me). So I figured I'd apply for academic jobs and if I didn't get one in a place that seemed ok, or if I got one but ended up having to spend too much time trying to get grants, then I would leave academia and figure out what to do instead.

Can you tell us a bit about what day-to-day life is like in your current position?

In normal (non-COVID) times, my day-to-day life was very very different from one day, or at least one week, to the next. That was by choice. I don't really like routine, and contrary to all the (probably very good) advice out there, I've never been able to develop a routine. So some days/weeks, I would have almost a 9-to-5 schedule, which typically involved a lot of one-on-one meetings, mostly with grad and undergrad students. On a day like that, I could often have five or six back-to-back meetings. I would often meet with grad students at a coffee shop or restaurant, and we'd have a tea/coffee, or sometimes even a meal, while meeting (always on me, at least in recent years). Undergrad meetings are usually in my office or lab. Other days, I might be teaching, in which case I often won't schedule anything before (to prep) or after (to recover – I am wiped out after teaching). And then there are editing days – those are usually on the weekend. I settle in at a comfortable coffee shop and get in the zone and read/edit manuscripts for 4 or 5 h without stopping. I love those days, but I tend to put them off, almost as a kind of treat or reward (which isn't really a great way to do editing, because often I'll sit down and see that a few of the papers I've been sitting on are very easy to deal with – I can make a decision about whether to send them out for review or desk reject them within 20–30 min). And then there are travel days. I used to have *a lot* of those. I said yes to almost any opportunity to travel. So I probably spent about a third of my days on the road. These trips were mostly to give talks at other universities and occasionally for conferences, grant panel meetings, or other kinds of meetings. That covers about 80–90% of what I spent my time doing. The rest is mostly replying to email and having more meetings.

What sort of role do you have with respect to journals and editing? I feel that PhD students tend to only have limited experience with that side of academia. What led you to become more involved with those types of roles?

I've had a few different roles with journals. In addition to reviewing for journals, I have also been an editor (associate editor or editor-in-chief) at a few journals. Each of those roles – reviewer, associate editor, and editor-in-chief – comes with increasingly more decision-making responsibilities and is increasingly more rewarding (for me – some people hate editing). Editing definitely isn't for everyone, and anyone who doesn't like it should by no means feel obligated to do it. What I like about it is that each manuscript is like a puzzle, and I learn so much from evaluating it and reading others' evaluations of it. I love teaching research methods and critical thinking, and to me, editing feels like critical thinking in action. And I feel like critical thinking is maybe the most important skill we can work on, and there is always more to learn. I've been editing for about 10 years, and I still love it and still feel like I learn so much. At the same time, seeing how journals work has made me very disillusioned about journals and peer review. It's weird to enjoy being part of a system that I think pretty much fails to achieve its objective – journals are supposed to be some kind of safeguard against bad science getting out there, but they are terrible at that. So I also spend a lot of my time and energy thinking about how to achieve that goal, probably outside of journals. There is no reason to do peer review the way journals do it – in private, through personal requests for favors, without much accountability or transparency.

How did I become involved in journals? I think probably just by accepting a lot of review requests as an assistant professor and then saying yes when I was asked to become an associate editor. The fact that there are very few tenured women in personality psych also meant that there were more opportunities for me earlier on than there might otherwise have been (as journals become more self-conscious about having diverse editorial teams). And then for the editor-in-chief positions, I applied and did not always get them! To be honest, I was pretty shocked to get as many opportunities as I've gotten and such good ones too. I think a lot of it was luck (as with many other things in my career).

Apart from luck, what are some factors that you think have helped you get more influence and opportunities in the field? You've mentioned both invited talks and editorial roles, and I'm sure those in tenure-track positions would appreciate your advice on getting out there.

It's hard to know for sure, but I'm happy to speculate! There are some factors that were out of my control (so maybe they count as luck), like having an advisor who made it a priority to give me opportunities to meet and work with other people and who talked me up to his colleagues and collaborators. That surely helped a lot. But in terms of things I had control over…. One is that even though I'm quite introverted in general, I have a habit of talking and asking questions in academic contexts. Asking questions at talks was probably one of the first ways I got to start having conversations with other people in my field. Then, the conversations started happening more informally, and even though I was pretty shy and quiet, I was able to have some really good exchanges with people at conferences, or when they would visit my department. Another factor is that I like to travel a lot and had the means to do so (including money and no dependents). For a while, as an assistant professor, I was in a long-distance relationship and would meet up with my partner in places where one of us had a conference or other work meeting, so I would find myself in random cities. If I knew someone in my field in that city, I'd sometimes reach out to them and ask if I could stop by their lab and say hi, and sometimes, that would turn into a talk (sometimes just to their lab, sometimes to the social/personality area).

I'm not sure, but I think those kinds of opportunities might have made a pretty big difference. In addition to strengthening my connections, it also gave me the opportunity to talk to and learn from lots of different people and get exposed to lots of different ideas. Then, later on in my career, I joined Twitter and some of the Facebook groups where discussions were happening, and that provided a completely different avenue for listening in on and participating in conversations. There are quite a few similarities between the kinds of opportunities I got from traveling around and the kinds of experiences I like having on Twitter, but of course, there are important differences too.

You recently co-founded the Society for the Improvement of Psychological Science (SIPS). Can you tell us about the responsibilities that went along with this and what was your motivation to be in this leadership position?

In 2015, I had been hearing and giving talks at conferences about the replicability crisis, open science, methodological reform, etc. Those conversations were starting to feel unproductive and circular – the same arguments and counterarguments coming up again and again, and it felt like allowing this to go on would mean giving up on making meaningful change. I was getting impatient and wanted a way for people who were ready to start working toward change to be able to get together and move forward, without having to go around and around debating whether or not things needed improving. So I reached out to Brian Nosek and asked if there was already a group or meeting like this, aimed at people who agreed things could be improved and were ready to start working toward improvement. That started our planning for what became SIPS. In 2016, Brian and I, together with a small group of researchers, put on the first SIPS conference and organized a formal society and executive committee. I became president of that society in 2017 and served as a member of the executive committee through 2020. To be honest, I didn't expect the little meeting we organized in 2016 to turn into anything nearly this big. There were 100 people at that first meeting, which already exceeded my expectations. At our last in-person meeting (in 2019), we had to cap registration at 525, and I think there was even more demand. I don't think I realized that I was taking on such a big leadership position when Brian and I started SIPS. Having served on other societies' executive committees, I really didn't want to create a society whose primary goal was self-promotion and self-preservation – I see too many decisions being made in order to bolster the financial health of a society, or increase its membership, etc. I really wanted to avoid SIPS becoming a society that makes decisions based on those kinds of considerations. So I've tried to maintain a pretty detached attitude toward it from the beginning – if it grows, great. If it becomes obsolete or gets taken over by a better society with similar goals, also totally fine. SIPS doesn't need to exist; it doesn't need to get ever bigger; it doesn't need to bring in money for the sake of bringing in money. I think that took a lot of the pressure off for me – I wasn't trying to build something huge, and if everyone stopped coming to our little conference, that was ok. Even so, there was a lot of responsibility – if you're going to have a society, you have to take care to make sure it's inclusive and equitable, that you protect people (e.g., by enforcing certain standard of conduct at the meeting, by finding a way to give early-career researchers a voice without burdening them will bullshit service, etc.). Navigating those decisions was tough, but luckily the other people making those decisions with me were incredibly committed, thoughtful, bright people. I learned a lot about how to manage an organization and a bit about how to be a leader. I definitely developed thicker skin – I think that the unexpectedly fast rise in our popularity understandably led to a lot of scrutiny and high expectations. I had to learn how to listen to my own conscience and know when I was doing my best, when I was getting defensive, when to listen harder, and when to stop listening (I know that sounds bad, but I really think it's an important feature of many leadership positions). I've basically completely stepped back from SIPS now (as I write this in January 2021), except for being the editor-in-chief of the journal that SIPS is affiliated with. I am really happy that I was able to let go, and I feel confident that it's in good hands, though again, I don't really care if it ceases to exist, if that's what's best for psychology. Right now I think it's still doing a lot of good for the field. One of the other exciting developments is that I'm seeing several other fields build sister societies to SIPS, and I feel like they are taking advantage of the lessons we learned and building on them to create their own version of SIPS in their fields. It's fun talking to the people leading those efforts and also a bit like rewatching your favorite show, where you know exactly what obstacle the protagonist is going to come up against but there's not much you can do to prevent it. But I think all of us, across fields, benefit a lot from each other's experiences and from knowing that we're not alone – I was lucky when I started SIPS that there were at least a dozen people just as committed to (obsessed with?) making this happen in psychology. From what I see in other fields, some of the people leading similar efforts don't have as much support.

What do you like most about your work?

Getting to work with amazing people (mostly my grad students). I am completely flabbergasted at the quality of people I've gotten to work with, in terms of their brilliance, creativity, and work ethic, but also what they're like as people. It is mind boggling to me that these people want to work with me. I also really like the independence, as I mentioned above.

And what do you like least about your work?

How cutthroat it is to get a job in academia. It's depressing. I don't know what to tell graduate students because I don't want to make them cynical, but the reality is quite bleak in terms of the chances of getting an academic job (if that's what they want, or at least want the option). I'm also very cynical about journals and rewards/incentives more generally, but that doesn't get me down as much (though of course it's related to why the job market is so bleak – not only are there very few jobs, but who gets those jobs isn't very correlated with the things you'd want it to be correlated with).

Do you think that someone who just finished a PhD might not realize that some aspects of being a professor and running a research lab are parts of the job?

I don't think there's anything totally unpredictable, but I think it's easy to underestimate just how many meetings and emails there are. It becomes impossible to keep up pretty quickly, so devising a strategy for dealing with both is important. I still haven't quite figured it out myself, but keeping those things from completely taking over your schedule is a challenge. Your time becomes less and less yours, and you have to protect it if you ever want to get around to the stuff that doesn't have any deadlines or anyone pestering you about it.

Another aspect I underestimated is how much of a responsibility mentoring is. We've all been on the advisee side of the advisor-advisee relationship, but I didn't realize what being on the advisor side would be like, and that also took me a long time to fully appreciate. The most striking thing about it to me is the heavy responsibility. One of the things I don't like about academia is how much power/influence advisors have over their graduate students. I was lucky to have a fantastic advisor, and I now realize that even relatively small things an advisor does or doesn't do can have a huge impact on their grad students' careers, well-being, motivation, etc. I find that responsibility very heavy sometimes, and I don't know how others manage it. I think we should change the system so that grad students' fates don't rest so heavily in just one person's hands, mostly for the sake of grad students of course, but it would also make the role less overwhelming for advisors (though of course part of the problem is that some of them want that much power/control). For me, one of the consequences of this system is that I think I can only be a good advisor to two or maybe three students at a time. I believe that other people can manage more, but I don't know how they do it. It's kind of disappointing because mentoring grad students is one of my favorite parts of the job, so having to restrict myself to just a couple students can be disappointing, but I think it's the responsible thing for me to do. And even with just a few students, I often feel like I'm not doing enough, or I'm holding them back. I don't think that's just in my head – I think it's often accurate, in part because many graduate students have such incredible skill and talent. It's a weird thing to live with day in and day out.

If someone currently finishing their PhD was considering a position similar to yours, how might they decide if it would be a good fit?

Luckily, I think that the grad school experience provides quite a bit of relevant info for people to decide if academia would be a good fit. There are some things that change when you become a prof, but a lot of the pros and cons are the same. I think for people who could imagine continuing doing similar things to what they were doing in grad school, pursuing a tenure-track position might be a good fit. I suspect in that case, the major new factor to consider is what the job market looks like and whether you'd have to make other sacrifices you don't want to make (e.g., living in a place you don't want to live, or having to make compromises with close others' careers, etc.). So if someone is finishing their PhD and thinks they would like a tenure-track position, my advice would be to apply (assuming the feedback you're getting suggests that your application would be competitive; if not, then consider applying for post-docs first), see what offers, if any, you get, and then decide if any of those offers are appealing enough to accept, all things (including personal life and preferences) considered.

If someone was interested in pursuing a similar career path, what would you suggest they do to better prepare themselves?

There are two ways to interpret this question. The first is what should someone do to have the strongest job application possible. The answer to that question, if they're interested in a faculty position at a research-intensive university (at least in the regions of the world I'm familiar with), is simple: submit papers for publication. Nothing else matters nearly as much as getting those manuscripts written up and submitted. In my second-to-last year of grad school, a few of my grad student friends and I realized that all of us were sitting on studies that we'd run, analyzed, and then hadn't written up. So we made a pact that we would try to submit one paper per month (of course, this is only semi-realistic if you're sitting on completed projects). I stuck to that goal for about 4 months, and that completely changed what my CV looked like when I went on the job market a year or so later. Those four papers made up a big chunk of my publications, and again, lots of luck was involved in getting those papers accepted in time for the job applications. One of them wasn't even a "real" paper – I wrote up a paper describing how I'd used a particular method (using the Internet to collect informant reports about participants' personalities – it was a pretty new idea back in 2004). For many of us, by the end of grad school, if you make time for writing up papers, you find that there are actually quite a few papers that you're in a good position to write, if you're creative about it. Of course, writing up papers just for the sake of writing up papers is one of the ways our field got into the replication crisis mess that we're in, so hopefully you're in a position where you have useful things to write up and share. But if the question is what an advanced grad student who wants a research/teaching faculty position should do, the answer is pretty clear given our current incentive/reward structure: write and submit papers.

The second way to interpret this question is what experiences should a graduate student seek out in order to be better prepared if/when they get a tenure-track position doing research and teaching (not what will help them get the job, but what will help them be well-prepared to do the job). There I think the answer is less clear. For me, grad school was the last time I had the freedom, time, energy, patience, etc., to learn new skills from scratch, so I would say try to learn any skills you think you might need (e.g., R, GitHub, SEM, multilevel modeling, etc.). Also, if you think you will have less freedom to travel/visit with people later on, do it now (if you can). Take every meeting you can get with other people whose work sounds interesting to you (unless you have a good reason not to, for example, you have good reason to think they're a jerk). Seek out opportunities to chat with people who do interesting work or have shared interests. I would be reluctant to make commitments to collaborate with people – that doesn't have to be (and in my opinion should rarely be) the outcome of networking – but just talking and having a basis for future communication can be very valuable. I have a very high bar for starting new collaborations for a couple reasons. One is that it's too easy to spread yourself too thin. The other is that I think it takes a long time to know if both parties have compatible working styles, values, expertise, etc. (Compatible doesn't have to mean the same, but not clashing.) Collaboration is a very intense relationship (most of the time), so I want to be sure that everyone knows what they're getting into and is comfortable with it, which is hard to do without getting pretty well acquainted first. Also, I think most of the time, not having enough collaborators is not the limiting factor on productivity/contributing to science. I think most of us have a lot we could contribute just on our own or with the people we already know we work well with, if we had the time, resources, etc. My last piece of advice about how to better prepare yourself is not wasting a lot of time on long-shot grant proposals. This kind of goes along with prioritizing writing up papers, but I also think if we all reflected on the time we spend writing grant proposals, it would be pretty clear that the vast majority of those proposals were not worth writing up. This likely varies a bit by subdiscipline, but if you don't absolutely need a grant to do your research, I would say only apply for grants that (1) would allow you to propose almost exactly what you want to be doing anyway and (2) that you have a decent chance of getting (i.e., the funding rates are not insanely low, and/or you are an especially good fit for the call). For my subdiscipline of social/personality psych, I think it's fine and common for people not to apply for grants at all until they are in a faculty position and then to only apply to one or two mechanisms/agencies that are a very close match to their research.

At the start of the interview, you mentioned that you recently moved universities, as well as countries. Can you tell us a bit more about that process? For instance, how long had you been looking for a new position, reasons for the move, or considerations for how universities might work differently in different countries (e.g., funding). Students hear about people getting their first faculty position, but moving from one faculty position to another is much less discussed.

I've actually moved twice since getting my first faculty position, so I'll talk a bit about both. The first was from Wash U (in St. Louis, USA) to UC Davis (in California, USA). That was a few years after tenure (in 2014, I started in 2007 and got tenure in 2012). The second move was this year (2020) from UC Davis to the University of Melbourne (Australia). I'm not sure how much generalizable knowledge can be gleaned from either of these moves, so feel free to take or leave any of this info.

The first time, I wasn't looking to move. I was happy at Wash U. I was spending a sabbatical year in California (at Stanford), and during that time, UC Davis was doing a search for an assistant professor in my field. A colleague there asked me who I knew of who was on the job market, and we discussed who in our field might be looking for an assistant-professor level job. As a result of those conversations, the question came up (actually through a mutual friend, not directly with the UC Davis person) about whether I would ever consider a job at UC Davis. I responded honestly that it's possible that I would, if it was a tenured position (since I already had tenure and was happy at my job, but also love California and being near my mom). That info got back to people at UC Davis, and they looked into whether there was a mechanism at UC Davis for a targeted search (many universities have these kinds of mechanisms, sometimes linked to diversity initiatives or other priorities). I was lucky that there was, so they invited me out for an interview using that mechanism. I remember that up until the interview, I felt kind of bad because I thought there was a slim chance I would actually end up accepting an offer even if I got one (I told friends honestly that the probability was around 20%). But then I got the offer and really wanted to go. I don't know what changed in my mind, but it was clear to me after the interview and offer that I wanted to accept. It certainly wasn't any desire to leave Wash U, but by that point, I'd been living back in California for most of a year, and it was very salient to me how much I liked being back there and being near my mom. It was still a difficult transition because of everything and everyone I left behind in St. Louis and at Wash U, not least my graduate students (one moved with me, but several stayed at Wash U).

The second time is kind of a crazy story. I visited Australia for the first time in November 2018. I was very happy at UC Davis, and if you'd asked me if anything could make me leave, I would have said with 98% certainty that the answer was no – I could see myself staying there my whole career. Two days after I arrived in Australia for my 3-week visit, I met my now partner. We'd interacted quite a bit before (online, via email, and almost but not quite at a happy hour at a conference), so we knew a bit about each other and had many mutual friends, which might help explain the next part, but really there is no good explanation. We got really serious really fast, so that when, 2 weeks later, after giving a talk at the University of Melbourne, I got an email from a headhunter asking if I'd be interested in applying for a professor position in their psych department, it felt not – 100% – crazy to consider it. I knew enough about psychology in Australia to know that, of all the universities in Australia, the University of Melbourne would be the best fit for what I do (both for personality psychology and for my metascience research). And this position sounded pretty perfect – it was a senior position they'd been trying to fill for a while, so it was a pretty cushy gig with a lot of freedom, great colleagues, etc. Still, I had to decide that week – the application deadline was a few days later. I decided to apply and see what happens. The rest is pretty typical job-interview stuff. I had a Skype interview first, then a fly-out (they also interviewed several other people), then another Skype interview, then the offer. By the time I had to decide whether to accept the offer, 6 months had passed and I had more confidence in my new relationship. I also was able to defer the start date for another year, which gave me time to wrap up some things in the USA and at Davis. Still, again, I was going to have to leave graduate students, colleagues, and friends behind, and that was not easy. I didn't and still don't know a lot about how the Australian system is similar to or different from the US system, so for that reason and many others, it was a bit of a leap of faith. However, through the interview and negotiation process, I got a sense of how the administrative side of things worked and also got a sense of the department culture, both of which seemed reasonable and kind of familiar. There are definitely some major differences (e.g., in Australia, academics rarely teach an entire course start to finish – almost everything is team taught, which is very strange to me, and grad school is only 3–4 years with no coursework, etc.), but many aspects of Australian culture and Australian academic culture seem quite similar to the USA. As much as I love my partner, I wouldn't have taken the leap if I didn't have a very good feeling about the move, and everything I saw in my visits and interactions with my future colleagues made me think it was a safe bet. In the end, I felt like I had two very good options: stay at UC Davis or move to the University of Melbourne, both of which were very appealing to me from a career standpoint and one of which – Melbourne – was much more appealing from a personal standpoint.

I don't know if I would have considered making the move earlier in my career. Factors that I think would have loomed large for me earlier in my career matter less to me now, like the prospect of being physically distant from so many of my colleagues and collaborators, of being overlooked for various roles I really want and enjoy, etc. I was lucky to get to do a lot of those things before my move, and I felt like I was in a good position to mitigate those negative consequences at this point in my career. It also helped a lot that I'd become active on Twitter and knew that I could stay connected and involved in global conversations that way. I still worried that no one would want to meet with me if it required doing so over videoconferencing rather than in person and that I'd miss all the conferences and invitations to visit other departments. Then the pandemic happened, and now everyone is videoconferencing and no one is visiting each other's campuses. Maybe if/when things go back to normal, I'll feel those losses, but I don't know yet.

Based on your journey, what advice or suggestions do you want to pass on to someone who's currently finishing their PhD?

I think one of the toughest challenges is figuring out how much you have to play the game, versus stick to your own sense of integrity. I was lucky that I was pretty oblivious to a lot of the threats to my integrity when I was earlier in my career, though not all of them, in part because that was before the replicability crisis and open science movement in psychology. But even if it weren't for the open science stuff, there would be other decisions we each have to make that force the issue – do you do what's best for your job prospects or what you think is right? I've faced that decision over and over again. I know the answer seems simple: do the right thing. But it's not always clear – look at something like reporting sexual harassment. There are many valid reasons for people not to come forward. The same kinds of issues around power dynamics, feeling like our attempts are futile, etc., come up in many different areas of the job. So I think one thing I've learned is that you have to know where your line is – where will you compromise and what's a dealbreaker? The other thing to think about is that this line should move as you get more power. Too many people become less likely to speak out, and more likely to toe the line, as they get more and more power and job security, but we should all push ourselves to move in the opposite direction. It's much more ok to make compromises when we're in precarious positions than when we've got power, so think about what compromises you're willing to make now and what you want future you to be more outspoken about. For me, that means thinking about the people I admire, and asking myself why I don't behave more like them, and then pushing myself to challenge the excuses I come up with.

Related to this, I would tell someone finishing their PhD that your skills are valuable and you likely have many options after your PhD (including quitting before you finish your PhD). So if you know your own limits or dealbreakers, you should feel empowered to walk away or rock the boat if you reach those limits, and know that there are plenty of paths to having a good, happy work life.

Thank you for doing this interview. It was great to discuss these topics with you!

"Running a Research Lab Is Like Running a Small Business with a Highly Uncertain, Constantly Fluctuating Budget"

Lucina Q. Uddin

Lucina Q. Uddin

Abstract In the interview with Lucina Uddin, we discuss how being open to opportunities can determine where you go next. Applying for faculty positions and deciding where to move to can be difficult, so it is important to think about your lifestyle preferences and deal breakers. Lucina highlights how working as teams and collaboratively is important and can be one of the best parts of working in academia. Unfortunately, there are issues in academia, such as those related to diversity. Academic roles change greatly as you shift toward leading a team—where skills such as emotional intelligence, open mindedness, decisiveness, and humility become increasingly important.

Contents

L. Q. Uddin (✉)
Psychiatry and Biobehavioral Sciences, University of California Los Angeles, Los Angeles, CA, USA
e-mail: lucina@ucla.edu

C. R. Madan (ed.), *Academia and the World Beyond*,
https://doi.org/10.1007/978-3-030-82606-2_22

Chris: Can you introduce yourself and tell me a bit about your current position?

Lucina: Sure. I was born in Bangladesh and immigrated to the United States with my parents before my first birthday. I guess technically they immigrated, and I came along for the ride. Growing up, at one point, I heard the silly myth that you only use 10% of your brain, and that enticed me to go into neuroscience. I am currently a professor of psychiatry and co-director of the Center for Cognitive Neuroscience Analysis Core in the Semel Institute for Neuroscience and Human Behavior at the University of California Los Angeles (UCLA).

What was the focus of your PhD?

It's funny, people always seem to think that they have to choose their PhD topic very carefully. I would say that the only really important decision to make when going to grad school is who will be your primary mentor. I didn't know what I was doing when I started grad school in 2001 in the Psychology Department at UCLA. I started my PhD straight after finishing an undergraduate degree in neuroscience, minoring in philosophy. The truth is I didn't want to go to medical school like all my friends were doing, and I didn't really know what else to do other than stay in school. So I applied to several graduate programs in neuroscience and didn't get into any of them. I did manage to get an interview in the neuroscience program at the institution where I had just completed my bachelor's degree, UCLA.

During the interview (which I generally bombed), I met Eran Zaidel who would come to be my most inspiring mentor in neuroscience. He eventually somehow convinced the Psychology Department at UCLA to admit me, even though they had already completed their graduate recruitment process for the year. I like to think it's because of a great conversation we had, where he asked me if I thought I could win the Nobel prize. I figured it was a trick question and said "I don't see why not." I entered his lab thinking (like many young neuroscientists) that I was going to study consciousness. I was intrigued by commissurotomy or "split-brain" patients, who have had their corpus callosum surgically severed as a treatment for intractable epilepsy. Roger Sperry (Eran Zaidel's advisor, who *did* win a Nobel prize) thought these individuals housed two separate consciousnesses in their disconnected cerebral hemispheres. So I spent the next 5 years conducting a series of behavioral and neuroimaging experiments on the neural basis of self-face recognition, the closest I could get to operationalizing consciousness at the time. Those were some of the best years of my life. It was the early 2000s, and we hadn't hit any financial crises yet. I had an NSF graduate research fellowship and a great mentor and worked in an extremely collaborative environment where I met outstanding scientists whom I'm happy to call friends to this day. Sadly, Eran Zaidel passed away during the summer of 2021. I like to think that he would be happy to know that I've since returned to UCLA as a faculty member and am carrying on his work.

As you were finishing your PhD, what were you thinking about your career plans?

Once again, I had no idea what to do after I completed my degree. I figured I would look for a postdoctoral fellowship, since that was what most people did coming out of a cognitive neuroscience graduate program in those days. I made the impulsive decision that I wanted to live in New York for a while, so I looked for positions advertised there. It was on the FSL listserv that I saw the ad that led me to my first postdoctoral position at the New York University Child Study Center, where I spent 2006–2008 learning about the brave new world of resting-state fMRI and human connectomics from F. Xavier Castellanos (another great mentor) and his group. While the work was very exciting and we felt like pioneers in those years, I'd be lying if I said I didn't start considering non-academic careers at that point. I knew the odds of obtaining a tenure-track faculty position were slim, and they were actually zero right after the global financial crisis of 2008. I went on to apply for faculty positions every year from 2008 to 2012. Yes, that's 5 years of going on the job market. In 2008, I applied for three positions and got no interviews. In 2009, I applied for 12 positions and got no interviews. In 2010 (I was actually teaching in Bangladesh at the time, which is another story), I applied for nine jobs and got no interviews. In 2011, I applied for over 30 positions and got three interviews, but no offers. Finally, in 2012, after applying for over 60 positions and going on seven interviews, I got five offers. And I was exhausted.

That's great that you finally got offers and multiple of them! Can you tell us what were some factors that helped you decide which to take?

I considered all the usual factors: money (start-up package and salary), collegiality of departmental colleagues, location of the university, etc. I also have to say that in the end, after making pro and con lists, consulting with friends and mentors, and agonizing about all the unknowns, I went with my gut. I just got a good feeling from the chair of the Psychology Department at the University of Miami (where I first started my lab), and the rest of the colleagues I met were so positive and enthusiastic. Going on a second visit to Miami in February after finishing a second visit to a wintry Chicago sealed the deal!

Can we go back to your other story. Why did you decide to teach in Bangladesh for part of 2010?

It all started with an email I nearly deleted without reading. In 2009, I was working at my second postdoctoral fellowship in Child Psychiatry at Stanford University. At the time, I had just obtained funding for a mentored career development award (K01) from NIH to examine large-scale brain networks in children with autism, and things were going well. On September 29, I received a mass email from the dean describing a call for applications for teaching fellowships at the Asian University for Women (AUW) in Bangladesh. Stanford had partnered with AUW to support teachers to spend some time over there and was running an exchange program with students from that university who would get to spend time over here. Now, I will be the first to tell you that I don't believe in fate or any higher power, but I nearly fell out of my chair when I saw that email. As a person from a country that many people have never heard of, you can imagine my surprise at receiving such an email, that seemed almost perfectly targeted to me. When I looked up AUW, I was further shocked to find that it was located in Chittagong, *the city where I was born*! It was just too much of a coincidence for me to ignore.

Of course, there were several complicating factors to consider. For one, I had just gotten the notice of award for my K, so I was supposed to be working on that. Also I was still actively and increasingly despondently on the job market, as previously discussed. I worked in the lab of Vinod Menon at the time, and when I mentioned the teaching opportunity to him, he suggested that it might be possible to delay the start of the grant for a few months, take a break from research, and accept the teaching job. Maybe he was so accommodating because he is also originally from South Asia, and he understood that I couldn't really ignore a chance to go back and serve in the country of my birth. Many countries in South Asia, especially third-world ones like Bangladesh, experience a "brain drain" whereby talented people want to leave the country to pursue academic and occupational opportunities not available locally. I think some part of the immigrant mentality always involves a feeling of guilt, of abandoning family and culture and home. I was too young when we moved to the United States to really understand the struggles my parents faced in those early years after arriving here. My father was a professor and founder of several colleges in Bangladesh. When he had the opportunity to pursue a PhD in the USA, of course, he jumped on it, even if it meant starting his career from scratch in his mid-30s after immigrating. I grew up with the idea that if I was ever presented with the opportunity to go back and really give back, I would do it. In countries with very corrupt politics like Bangladesh, foreign aid in the form of money often gets diverted to the pockets of politicians and never reaches its intended target. So, all in all, I thought that if I could physically go there and teach young women in a classroom, that would be something that could not be taken away from them, or from me.

Can you tell us a bit about what day-to-day life is like in your current position?

Pre-COVID, day-to-day life involved a lot of travel. I'm not proud to admit that my carbon footprint was large: I was probably getting on a plane two to three times per month. It's hard to say no to invitations to speak at conferences, give departmental colloquia, serve on study sections, or basically anything else when you are an early-career researcher. I think I still have the mentality that I have to say yes to everything in the off chance that it could lead to a career-enhancing opportunity. So yes, I was traveling a lot. Nowadays, I've welcomed the opportunity to stay put for a few months, avoid jet lag, and catch up on reading and writing (both academic and non-academic).

A typical day at the university would involve meetings, meetings, and then more meetings. I guess that's one reason I enjoyed traveling so much, to have an excuse to miss them!

Do you think that someone who just finished a PhD might not be aware of all the different aspects of the job of professor?

Running a research lab is like running a small business with a highly uncertain, constantly fluctuating budget. I've always subscribed to and been a huge fan of *Harvard Business Review*, which consistently provides great articles on management, leadership, motivation, and practical matters when it comes to business.

The other thing I have found interesting to note is that the skills that make you a great PhD student or postdoc do not necessarily overlap with those necessary to run a successful research lab. As a trainee, your technical skills (e.g., coding) and in-depth disciplinary knowledge are constantly evaluated and critically contribute to your career progression. As a PI, you will mostly be concerned with two things: writing grants (that get funded) and writing papers (that get published). To do those things, you'll need to successfully motivate a team of individuals with diverse skills and backgrounds. I think it has something to do with the quote: "If you want to build a ship, don't drum up the men to gather wood, divide the work, and give orders. Instead, teach them to yearn for the vast and endless sea." I see my job as a PI as primarily to channel the motivations and talents of young, enthusiastic scientists into tangible, impactful contributions to science. This involves a lot of delegation and mutual trust among lab members, which you as the PI must cultivate. The skills that I think are therefore most critical in a PI are emotional intelligence, open-mindedness, decisiveness, creativity, and a high tolerance of uncertainty. And of course, a healthy dose of humility and the ability to learn and quickly bounce back from rejection.

If someone currently finishing their PhD was considering a similar position as you have now, how might they decide if it would be a good fit?

I think many people are surprised (either pleasantly or unpleasantly) to find that being a PI is very different from being a trainee in someone's lab. To decide whether this might be a good fit for you, consider whether you enjoy and are good at doing the things that are expected in the role including writing, fundraising, traveling, giving presentations, mentoring, science communication, training the next generation of scientists, and teaching courses. It turns out I personally love doing most of those things, but you have to be honest with yourself about whether or not a lifetime of those activities will suit you. I always encourage my students to seek out summer internships in industry to explore possible options outside of academia, and as a result, many of my former trainees have found success in data science and user experience research (getting paid much better than I do as a professor!).

If someone was interested in pursuing a similar academic career path, what would you suggest they do to better prepare themselves?

First, set realistic expectations and goals based on your true motivations and lifestyle preferences. I always say that to stay in academia, you will probably get to choose two out of three of the following (at best): (1) a job/career/position you really like, (2) living in the city/part of the world you really want to live in, and (3) living near the person/people (e.g., significant other, family) you want in your life. It's really hard to go for all three with the constraints of the academic job hunt, so figure out what is a deal breaker for you.

Once you have mentally prepared yourself, the hard work of building up an impressive CV begins. Although papers and grants count for a lot in academic settings and everyone knows that, I think it's also important to establish yourself as a team player, someone who others see as a competent collaborator and good mentor, an enthusiastic champion of open science and diversity, and in general a "good sport." No one wants to hire an asshole.

Of course, even doing all the right things is no guarantee that the well-deserved role will be available when you wish to accept it. See above (set realistic expectations), and have a plan B. If I weren't a scientist running my own lab, I would probably be working in science communication or journalism, perhaps with a side stint as a yoga instructor or high-school cross country coach. If I were independently wealthy, I'd be writing science fiction novels or maybe taking a stab at stand-up comedy (I did it once and it was fun!).

Do you have any suggestions on how to balance being a collaborator and team player versus establishing yourself as an independent scientist?

When you are first starting out in academia, you might be coached that you should avoid the "collaboration penalty." That is, in faculty hiring and promotion, sometimes scientists who are seen as merely collaborators rather than independent generators of new knowledge get penalized. This is an unfortunate old school mentality that glorifies the lone scientist, who in reality does not exist. We all work within teams—the only difference between those seen as "merely" collaborators and those viewed as solitary geniuses is the credit that the rest of the team gets for their contributions. Personally, when I see that a PI is the sole author of an empirical neuroscience manuscript, I often wonder whether that PI *really* single-handedly conceived of the project, collected all the data, conducted all the analysis, created all the figures and tables, and wrote and revised the manuscript. If they did, good for them! But more often than not, I see PIs who do not give appropriate credit (in the form of authorship) to trainees who clearly contributed to a project. I prefer to give credit where credit is due, and when in doubt, err on the side of including more authors to fairly acknowledge everyone's contribution. Did an undergrad collect data that went into the paper? If so, in my opinion, that undergrad should be listed as co-author.

I love collaborating. The majority of my publications in recent years have been co-authored with scientists from around the world. My collaboration network is large and continuously growing (see below). I prefer to be involved in team science whenever possible, as I find it is the best way to conduct the most robust, impactful, interdisciplinary research. Over the years I have been involved in a variety of funded and unfunded research collaborations and have contributed to teams of varying sizes including individuals with backgrounds in psychology, neuroscience, computer science, physics, medicine, electrical engineering, and philosophy.

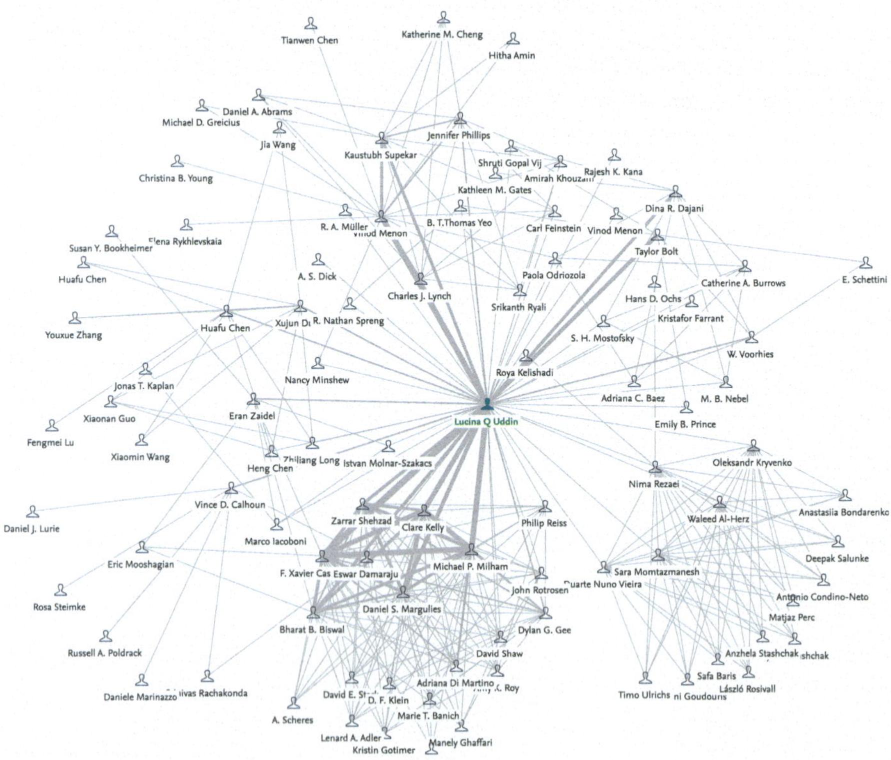

That being said, there is still the need to establish one's independence and demonstrate which aspects of your research program are uniquely yours and which are team efforts. To do this, I would advise writing sole author reviews every once in a while to establish your voice and your niche. This is especially important for women, who are often given less credit in collaborative projects due to gender biases that are entrenched in our society. When you are the sole author on a review in a journal that is well regarded by your community, there is no question that the ideas came from you.

What do you like most about your work?

The opportunity to engage in exciting collaborations (see above) and travel! I was probably traveling way too much before the pandemic, but I always enjoy exploring parts of the world for the first time and being exposed to new cultures and foods. I also love the flexibility that comes with academia. Last year I was on sabbatical. What other profession pays you to take a break from your daily work routines and just think for a whole year (every 6 years)?

I also truly enjoy mentoring others, and I guess I'm not terrible at it—in 2020, I won the University of Miami Department of Psychology Outstanding Faculty Mentoring Award. I think mentoring is the most important thing we can do as scientists. Sometimes, it just takes a little bit of encouragement, sharing of resources, and networking to get to the next level, and it always makes me happy when I'm able to do that for up-and-coming researchers. In addition, I always learn a great deal from my mentees; it really is a mutually beneficial thing.

What do you like least about your work?

The fact that academia is not a pure meritocracy, no matter how much people insist that it is. Since there are so few faculty positions available, biases always come into play when decisions are made regarding who will get those coveted positions. The same goes for successfully applying for grants, publishing in high-impact journals, getting speaker invitations, and everything else where there are gatekeepers. And guess who always seems to get the short end of the stick whenever gender and racial biases come into play? Just like the rest of society, academia is biased and structured in such a way so as to perpetuate inequalities.

One thing that I have found particularly exhausting over my 20 or so years in academia has been the extra work and expectations placed on women of color. There are three main ways in which this manifests: (1) We have to be at least twice as productive (scientifically) as our non-minority colleagues to even be taken seriously as a scholar in the first place; (2) we feel personally obligated to informally mentor and in other ways assist other under-represented junior scientists who sometimes see us as their only allies; and (3) we are expected to formally participate in (or in some cases create, maintain, implement, or expand) diversity initiatives every time the establishment decides it is in their best interest to respond to a given social justice issue.

I think this is an important ongoing issue in academia. Do you have any suggestions for websites/resources that might be helpful in giving others more direction in how they can help contribute to diversity initiatives?

There are so many resources out there now that I don't think the problem is that people are unaware of these issues. The problem is that a large proportion of those with the power to enact antiracist policies do not wield that power. I'm a big fan of Toni Morrison who said "If you have some power, then your job is to empower somebody else." I've made it a personal mission to use every bit of power conferred to me by my positions to do exactly that. I lead my own lab in such a way so as to empower trainees from diverse backgrounds to find their scientific voices. I try to engage with every committee, society, and editorial board that I am a member of with the same philosophy.

I think in general the OHBM Diversity and Inclusivity Committee has done great work in the past few years toward embracing diversity initiatives (https://www.ohbmbrainmappingblog.com/blog/seeds-of-change-within-ohbm-three-years-of-work-addressing-inclusivity-and-diversity). As the program chair for OHBM 2019 (back when we still had in-person conferences), I worked with the program committee to ensure gender and geographic diversity among the speakers and organized the first diversity symposium. As a senior editor of several journals, I try to balance the reviewers I invite so that they are not always from the same gender, country, or career stage. I recently published an opinion article on this topic with a colleague, entitled "Revising evaluation metrics in academia to dismantle privilege."

This great post by Sade Abiodun on Twitter makes it clear that these types of revamps are long overdue.

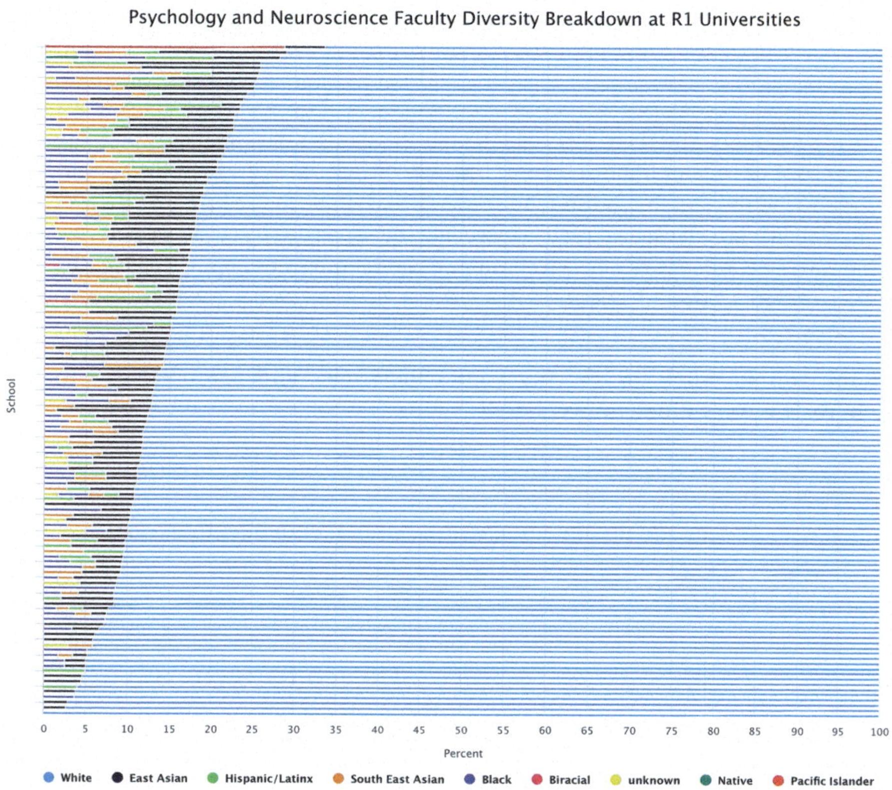

You see what I did there? I used my platform to draw attention to an important social justice issue, because I have the power to do so :).

Based on your journey, what is some advice or suggestions you would want to pass on to someone who's currently finishing their PhD?

If you are reading this in 2022, these are unprecedented times. If you are reading this in the future, I hope you will say "Wow, that was the worst ever, but now things are fine!" Go easy on yourself. If you don't have a concrete plan, who can blame you? It's hard to predict what new obstacles the next month will bring, much less the next 5 years. Accept that there will likely be some delays to accomplishing your life goals. Sometimes, unexpected insights come from forced idleness.

Is there anything else you'd like to tell someone reading this interview?

Forget everything you have just read! The truth about advice is that it only applies if your own personal circumstances align very closely with those of the person giving advice. I find that this is rarely the case. Still, I think we all love to tell the origin stories about ourselves, so I really appreciate the opportunity to do so here. If anything I've said here rings a bell for you personally, then I'm happy to have reached you. Now go forth and manifest the version of your career that brings you the least misery!

Thank you so much for sharing your experiences and perspectives with us here. It is very much appreciated!

“There Are Many Reasons That People Succeed in Academia”

Jamie L. Hanson

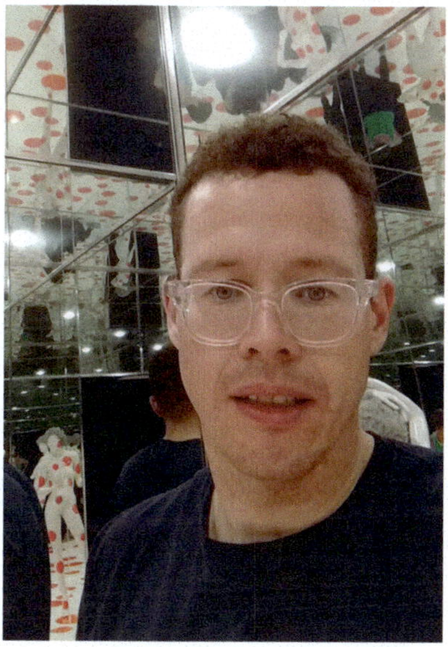

Abstract In our interview with Jamie Hanson, he tells us how he progressed from graduate school to a tenure-track faculty position and some of the challenges he encountered along the way. Jamie highlights that academic positions involve many responsibilities beyond research and teaching, something many PhD students may not realize. Of particular importance, many academic positions involve a shift from focusing on doing the research yourself to being in a more mentorship role. Jamie also shares some insights from considering positions at several universities and considering what might be best as a family, especially in light of a two-body problem. Academia requires a high degree of resilience to rejection, but there are still many paths to success.

Jamie L. Hanson

Contents

J. L. Hanson (✉)
Department of Psychology, University of Pittsburgh, Pittsburgh, PA, USA

Learning Research & Development Center, University of Pittsburgh, Pittsburgh, PA, USA
e-mail: jamie.hanson@pitt.edu

C. R. Madan (ed.), *Academia and the World Beyond*,
https://doi.org/10.1007/978-3-030-82606-2_23

Chris: Can you introduce yourself and tell me a bit about your current position?

Jamie: I am currently in my mid-30s and a father to a high-energy toddler, and I live in Pittsburgh. I love dancing and often describe myself as a "music nerd"; however, both of those things were more central to my identity pre-parenthood and pre-COVID. I am an assistant professor in the Department of Psychology at the University of Pittsburgh and a research scientist at the Learning, Research & Development Center also at the University of Pittsburgh. If we are getting in the "academic weeds," my position is "hard money," and I teach ~3 classes a year. My department is reasonably sized, with ~40 tenure-stream faculty and a number of admin positions, and in a school of arts and sciences at my university. Psychology is one of the more popular majors, so we serve a few thousand students at Pitt, including both majors and non-majors.

What was the focus of your PhD?

To introduce a theme that will likely pervade lots of my answers, I tend to live and focus in multiple different worlds. I was dual-mentored by one supervisor who specialized in neuroimaging and one who specialized in developmental psychology; both broadly worked in affective neuroscience. My actual PhD was in a "make your major" program at the University of Wisconsin-Madison. It might be more fun to say "choose your own adventure." Though it was really called an "individual graduate major" or IGM. This program is rather unique, but basically I selected a small committee of faculty to guide me through graduate school. I consulted with them on specific classes to take, what qualifying or "prelim," exams would be, etc. My committee consisted of my PhD mentors and two faculty members who conducted research in nonhuman animals, specifically rhesus macaques. I also had a good bit of informal mentorship from a psychoneuroendocrinologist who was originally a postdoc in one of my labs, a biostatistician specializing in technical neuroimaging analysis, and a health economist studying the impacts of poverty. As such, I often describe my degree program as one "that integrated developmental psychology, the neuroscience of emotion, public policy, and biostatistics." I think that's mostly accurate given all the other things I wrote here. I was in graduate school for a *long* time. I started in 2005, finishing up my degree right at the end of 2013, and I think I technically "officially" graduated in 2014.

As you were finishing your PhD, what were you thinking about your career plans?

I will be honest and say I am a bit of an "odd duck"; by the time I got to graduate school, my major career goal was "to be a professor." Now being a professor, I can say that I didn't actually know all the things that "being a professor" entailed. I, however, know that this was the right decision for me and, as a first year PhD, that I wasn't completely off-base with longer-term plans. Of note, again being an "odd duck," I went into college being a declared psychology major, taking as many classes as I could. A good random example of this, as a senior in college, I remember meeting with Dr. Robert Rescorla. Rescorla was an incredibly important figure in psychology and a pioneer in the study of learning; however, we met and I said I *just* wanted to take more and more psychology classes. He suggested I diversify and take a photography class or something in anthropology. I didn't listen to his advice.

Can you tell us a bit about your journey from finishing your PhD to where you are now?

Like most, the journey was filled with lots of ups and downs. Before finishing my PhD, I had a reasonable bit of success with research. I had 10+ published or in-press articles before my dissertation. This opened up a number of doors for me, and I was lucky enough to basically be recruited for a postdoc at Duke University; multiple labs pooled in funds, flew me down to North Carolina, and I interviewed with each one. I started with the incredibly supportive Dr. Ahmad Hariri, funded by a training grant out of the Center for Developmental Science at UNC-Chapel Hill. Working with Ahmad was great, but after the first year of my postdoc, the training grant didn't get renewed. Ahmad basically met with me and said: "I'm supporting you and another postdoc, and that other postdoc is more closely aligned with my grants. You should start thinking about finding other funding." I was stressed out and had to basically go around to different labs, some of which I didn't choose when I first interviewed at Duke, to find new funding support. That eventually worked itself out, and I was able to have a joint/split postdoc with Dr. Hariri and Dr. Kenneth Dodge.

Dr. Dodge was a developmental psychologist, completing some really seminal work in the 1980s and 1990s. He then moved toward larger public-policy foci, so it was a different approach and style from when I started at Duke.

With that crisis averted, I put my head down and kept trying to be productive on my postdoc. I submitted a few more papers and then hit the job market. I was again very fortunate to land a few interviews—I think maybe seven? However, it should be noted that I applied for ~30 jobs. It was then fairly grueling 6 months of visits and interviews. Of note, my partner was also an academic, actually on a tenure-track at that time. So my professional life and personal life were deeply fused at that time. She would tell me about her professional challenges, I would talk about a job I was applying for or a department that I was interviewing with, etc. There was little to no "space" and time for us to just be a couple. We both really hated the whole process. I would get excited about a job; she would have mixed feelings about that city or university. She would get excited, and then I wouldn't get an interview. We really tried to operate collaboratively and as a team. It was some of the best advice I have gotten about solving "two-body problems"—basically, make sure everyone is pretty happy with the decisions. If someone goes in with some hesitation or potential resentment, it will only grow and grow. Marriages and partnerships can basically be poisoned by that stress and those negative feelings. Eventually, again through lots of tears and tough conversations, we decided to leave North Carolina and go to Pittsburgh. My partner was sad to leave her university, but we thought it was the best decision for our family. Things mostly worked out. My partner ended up having a visiting professor job for a few years, but it was kind of a terrible match for her; that was really sad and frustrating. She liked academia but had some hesitation about all the facets of academia.

So I kept plugging away as an assistant professor. I had graduate students and undergraduates join my lab, my family had a baby early on while we were in Pittsburgh, and we have tried to survive all the different challenges thrown at us (getting papers and grants rejected, the pandemic, etc.).

From interviewing at different places, did you get an indication of how they differed? Apart from the two-body problem, what are some factors that an applicant might want to consider when deciding where might be a better fit?

Sort of. Nearly every place said they "valued collaboration," "appreciated interdisciplinary work," and "didn't want to be siloed." However, it was clear that in a few places, that was more just "talk." When places struggled to answer questions about faculty working across multiple areas in a department or outside of a department, that was an easy tell. I really, really value interdisciplinary work—my PhD in an individual graduate program that I designed, which was "area agnostic"—so I was happy to land where I did.

In terms of other pieces to consider, there are a few in-the-weeds things that are useful, if folks are considering multiple places. First, are there internal grants potentially available for junior faculty or more novel (high-risk) projects? For example, the research center I work in has a very generous internal grants program, as does the broader University of Pittsburgh. I think thus far, I have received about $200K in internal grants. The applicant pool is much smaller than an external grant review, so it is a major advantage. Second, when does the tenure count "start"? A few colleagues are at institutions (e.g., UC Davis) that count "everything post PhD" in terms of publications, grants, etc. My institution does not really do this. Third, does that department "actually tenure" people? That sounds a bit silly, but some departments are notoriously tough to receive tenure in and to rise up the academic hierarchy.

Finally, and really most important of all—if you are thinking about an academic job and have the luxury—consider the actual place you are living. Might it be a good place for someone to live? To relocate to? To raise a family? To buy a house? Basically, where you might want to spend 30 years, as that happens with many academic jobs, you go to a department and stay for a long time. A good example of this, of a colleague and university that I will not name—someone I knew who did their PhD in a major cosmopolitan city and their first job was in a very small town where a major research university was. While you sometimes can't control where positions are, etc., I personally couldn't see moving to a very small town where I don't think I could be happy or to a department where tenure is not very likely and a person is likely to move again in ~6 years when tenure doesn't happen. My colleague actually was very productive and would have received tenure but moved because I don't think the town was a reasonable, long-term location. *But* all that is to say—consider the psychological side of things—do department and the cities/towns they are nested in seem like a place you want to spend a few decades. That's kind of a tough question—like agreeing to marry someone after one date—but it is a reasonable analog. So do some fact findings and ask around; ask colleagues and professional friends for experiences and details of those surroundings.

Do you think that someone who just finished a PhD might not realize that some aspects of being a professor and running a research lab are parts of the job?

The first thing that really jumps to mind is *service*. I always thought professors just research and teach. If you were good at those two things, awesome*; but* there's this third major thing that dominates lots of your time, aka "service." This includes being on student committees, having faculty meetings, serving on a search committee for new faculty members, etc. Don't get me wrong—I love many of those tasks and elements, but you invest a ton of time in all of that service. There are days where I don't get much research done because I'm running from one service-related meeting to another, and then I have to answer urgent student emails, etc.

The other thing I would say is that every department is different and it's hard to define "success." Few if any departments will say you need "X" number of papers to receive tenure. And every department will conceptualize and count publications and other research accomplishments in different ways. Some universities will say "Oh you published 40+ articles; that's great!" Others will say "Oh you published a good bit, but most of them were with your old mentors and supervisors. It would be great if you published more independently." The lack of clarity can be stressful, as one always wonders "Am I doing enough? Is my position 'safe'?". Those things are really specific, but it's wild to be worrying about hitting some benchmarks, instead of just "doing science".

Can you tell us a bit about what day-to-day life is like in your current position?

My day-to-day schedule is reasonably flexible but often gets packed with meetings. There's a reasonable amount of "service" to my department or research center, and this is much more I think than folks realize. In a typical week, I would say there's 2–5 h of area group meetings, center meetings, student defenses, etc. This is all on top of the other meetings I have to support my graduate students and staff. You add in teaching there, my lab's meeting, and fairly quickly my open time evaporates. Most have likely said this before, but the amount of time I have to "actually do science" and not just write grants to do science gets *real* low, real quick. As a good professional friend, Jamil (Zaki), said once, "Imagine being an amazing college basketball player, and then all of sudden, you are transported to the NBA … but as a coach." I participate in lots of meetings and plannings about science but don't execute as much science as I did as a graduate student or postdoc.

What do you like most about your work?

One broad answer—the flexibility. I love that I have choice in the exact hours I work, which is especially nice and useful as I have a toddler at home. It is helpful when she is sick or childcare is unreliable to be able to focus on critical personal and family issues when needed. I also deeply appreciate the flexibility in what I research. Some weeks I get incredibly excited about a new method or idea, and I feel so lucky to have the freedom—"OK, let me figure out how to implement X or to focus on Y…." I really can drive what I am thinking about and reading. That feels unlike most any type of job outside of academia and research.

What do you like least about your work?

The meetings!? That's a semi-joke, but there is a reasonable amount of "service" related to being a professor—being on this committee or meeting about this issue. It takes a good bit of my time, almost every week. Sometimes, the meetings and topics are interesting and important, other times as the joke goes "it could have been an email…."

If someone was interested in pursuing a similar career path, what would you suggest they do to better prepare themselves?

I've said this before in different social media spaces and conference panels, but I think it is very true—PREPARE FOR REJECTION. This job is built on negative feedback, critiques, and reviews. Develop a thick skin and get ready to be resilient; get ready to fight for the science and work you think is important.

I think a lot about the advice that people give and sometimes the challenges I have had with mentors. One of my favorite people in academia is Dr. Andrea Hussong at UNC-Chapel Hill. Andrea has an uncanny knack to think and look outside of herself and her experiences, to truly connect with mentees, and to think about multiple "points" (or frames) of advice. In my experiences, many mentors often give advice and guidance from *their* vantage point, what worked well for them—basically their "N of 1" experiences. That makes sense but can be hard, especially as a new trainee. Perhaps you don't want to (*or can't!*) follow a mentor's path, so getting advice related to that isn't always the most useful. Related to this, I remember I had a mentor who always pushed me to do science very similar to their work. I, however, knew that wasn't my strength, that wasn't how I was going to make an impact; that mentor was very kind, but it initially frustrated me. I felt like I was doing science wrong and was of lesser value. It took me a bit to realize the science was just of a "different" value. The work they did and the approach they employed were useful, but so was mine. This felt like one of those "N of 1" moments. But it is important to think that, inside and outside academia, success comes through many paths—some straightforward, some roundabout, etc.

Based on your journey, what advice or suggestions do you want to pass on to someone who's currently finishing their PhD?

I guess my best advice is to think about your "end points," but be mindful of the realities of different situations. For example, graduate school is a good choice for many folks but isn't for everyone. It is an amazing space to learn and grow; a colleague once called it the "worst paid, but also most intellectually interesting, startup." You can pursue some amazingly interesting, new ideas, but, if you are going to graduate school, purely to "be a professor", I would suggest thinking some more. The number of academic jobs is *really* low and likely going to be even lower post-COVID (at least for a few years). You can, however, get amazing jobs in industry or other places that often pay well and allow for better work/life balance. Post PhD, getting one of those types of jobs is definitely a "win" in my opinion.

Other than that, I would say figure out how to deal with failure and setbacks because there will be *lots* of them!

Is there anything else you'd like to tell someone reading this interview?

This is an odd way to close things out, but assessing myself in the start of graduate school, I wasn't the smartest in my cohort by a long shot. There were so many capable and talented people I was studying with. I feel like I have been successful, thus far, and a great deal of that feels due to luck and "non-cognitive"/soft skills (e.g., persistence, dealing with failure, strategy).

I think this is a great point—not to say you're not smart, but rather that soft skills are very relevant for success. Can you tell us more about your thoughts on this?

I feel like there are many reasons that people succeed in academia—some people are just brilliant (Karl Deisseroth and Danielle Bassett types). Maybe you're brilliant too? But if not, there's not much you can do; other people are good at positioning their work, "selling it" a bit, and being hard and smart workers. They "grind it out" by leveraging all the other elements of their personality and skills. This includes folks who are diligent and persistent and can robustly deal with failure; it also includes really amazing managers/supervisors/mentors. And this isn't to say that the brilliant folks aren't good managers. These elements, which aren't exactly about "pure g"/intellect, are major drivers of success. Last thing is about being an interesting, or maybe even "fun," collaborator. I feel lucky—I've worked with some incredibly smart people, who were amazing writers, thinkers, etc.; however, some were more difficult to deal with than others. Being a scientist that someone "wants to work with" is also important. Science nowadays is a team sport. If someone is difficult to be around, it often leads to them being less in demand to work with.

Thank you for sharing your experiences, Jamie! It was great to hear about your path and insights.

Index

C. R. Madan (ed.), *Academia and the World Beyond*,
https://doi.org/10.1007/978-3-030-82606-2

The manufacturer's authorised representative in the EU is Springer Nature Customer Service Centre GmbH, Europaplatz 3, 69115 Heidelberg, Germany. If you have any concerns regarding our products, please contact ProductSafety@springernature.com

Printed and bound by CPI Group (UK) Ltd, Croydon, CR0 4YY
30/09/2026
02225232-0001